ENERGY
SPRAWL
SOLUTIONS

ENERGY
SPRAWL
SOLUTIONS

Balancing Global Development and Conservation

Edited by
Joseph M. Kiesecker and David E. Naugle

Foreword by
Peter Kareiva

ISLANDPRESS

Washington | Covelo | London

Library of Congress Control Number: 2016957966

Printed on recycled, acid-free paper ⊛

Manufactured in the United States of America
10 9 8 7 6 5 4 3 2 1

Keywords: Island Press, energy sprawl, Development by Design, hydropower, wind energy, solar energy, renewable energy, energy planning, conventional oil and gas development, offshore oil and gas development, marine spatial planning, fracking, coal mining, petroleum, natural gas, biofuels, energy access, energy equity, energy consumption, carbon dioxide emissions, spatial footprints, energy impacts, cumulative impact assessment, spatial optimization, energy financing, climate change mitigation, greenhouse gas emissions, international energy agency, multilateral development banks, world energy outlook, United Nations Convention on Climate Change

Contents

Foreword

This is a book about development by design. If you do not know what that is, then you have not been paying attention to conservation over the last decade. Development by design is a simple yet revolutionary idea for conservation. The authors in the book project the magnitude of our demands for energy, as well as the specific places where that demand is likely to be met—via solar farms, windmills, offshore oil, hydropower, biofuels, and so on. These are not data or model projections conservationists typically pursue. Our authors then draw on their deep knowledge of where lands and waters of the greatest conservation value occur, and they seek some way, however challenging, to accommodate both energy development and the protection of our planet's biodiversity. In other words they jointly consider biodiversity and energy.

Why is this necessary? Over a billion people worldwide do not have access to electricity today. That is wrong—it means no refrigeration, no lights for children to read by, businesses cannot run except in daylight, health services are restricted, and cooking is often done with firewood. By 2050, there will be between two and three billion more people on our planet than there are today—they will all need food, housing, and electricity. These are defiant trends that mean, one way or another, that our lands, rivers, and oceans will have to accommodate enormous growth in energy production. Much, ideally most, of the new energy will come from solar, wind, hydro, and nuclear power. While renewables have the greatest environmental appeal, even they can impact wildlife and biodiversity. Birds die in windmill blades and fields of solar panels are land hogs. Hydropower interferes with fish migration and distorts natural flood regimes. The transmission lines required to link distributed renewable energy into our power grid can fragment habitats and become conduits for nonnative species.

What then are our options? One option is to curb demand, by somehow decoupling human development and well-being from cheap energy. This could be done by increased efficiency, a circular economy that minimizes waste and reuses materials, and fundamental changes in consumer behavior. If at the same time technologies for energy storage and transmission could be improved, the pressure for massive new energy developments would be reduced. But even the most optimistic projections of social change and energy innovation cannot halt the short-term need for the massive energy developments that are already part of national or regional development plans. For example, California seeks to generate 50 percent of its electric power from renewables by 2030—a goal that cannot be achieved without massive new solar and wind installations.

If we are activists who are attached to special places, or conservation biologists who care deeply about biodiversity hotspots and critical population enclaves, we can engage in litigation and block specific energy projects. But curtailing energy impacts one litigation at a time is costly, creates social and political resentment, and does not solve the problem of needing more energy.

Alternatively, we can care as much about delivering energy as we do about biodiversity and adopt what has come to be called "development by design." In particular, we can use data and modeling to identify open zones for energy development and, conversely, "do not touch" zones for biodiversity protection. And we can apply engineering and ecological savvy to mitigate impacts on wildlife and biodiversity in those places where we do invest in energy infrastructure.

The hypothesis underlying development by design is that by identifying places where energy development can occur, one gains bargaining chips that will allow us to secure biodiversity havens and "do not touch" zones. A corollary of this hypothesis is that only by addressing infrastructure at regional and large landscape scales can one do the sort of whole-systems analysis that avoids death by a thousand cuts. Another corollary of this hypothesis is that an uncompromising draw-the-line-in-the-sand approach marginalizes conservation efforts from the bulk of the public and most decisions.

Of course, pragmatism has its risks. Development by design could be the Faustian deal that dooms our planet because we did not have the courage to take a heroic stand against development and energy growth. Debates about pragmatism and compromise versus unyielding stands against development are imbued with personal philosophies, politics, and values. But science and evidence informs these debates. For example, several case studies in this book document how a mix of stakeholder engagement, strong science, and a willingness to compromise can yield large new protected areas for conservation while also expanding a nation's energy supply. Of course, a handful of case studies do not win the argument. One could equally find case studies in which some massive energy development was allowed in exchange for an offset deal that is more public relations than a biodiversity gain.

We need to remind ourselves that we are trying to solve three big, hairy problems at once: (1) climate change and hence the need for a decarbonized economy; (2) cheap energy access for everyone; and (3) protecting habitat and wildlife even as the move to renewables fragments landscapes and uses massive amounts of land. Each of these problems is a mix of politics and science. The science underpinnings for a development-by-design approach requires that we answer several questions for which we do not have adequate methods or theory at this point in time:

- Can we identify thresholds of development and degradation that, if crossed, will cause irreversible loss of wildlife and biodiversity?

- Are there enough locations of limited conservation value in which energy development can be located?

- Is it possible to design an energy development project that results in no net biodiversity loss?

- What is the shape of the trade-off curve between loss of conservation value and delivery of energy? And how is this trade-off altered by optimal design and mitigation?

This book is about data and analysis and policies aimed at questions like those above.

The idea of balance is core to this book—how might we balance energy needs and conservation. *Balance* is an interesting word. To some it likely signals surrender. For that reason, conservationists tend to talk about win-wins (which eliminates any idea of surrender). But win-wins are scarce and trade-offs are stubbornly common. Much depends on which lands, or rivers, or waters we give over to energy development. Those decisions will determine what biodiversity we give up when we develop new energy sources, so that we can then somehow protect biodiversity in the long run while at the same time delivering cheap energy, and reducing greenhouse gas emissions. There is nothing more fundamental to conservation in the twenty-first century than wrestling with this trio of challenges, which is exactly what this book does.

Peter Kareiva
Pritzker Distinguished Professor in Environment and Sustainability at
    the University of California, Los Angeles
Los Angeles, CA

Preface

In the well-known words of Thomas Friedman, we live on a hot, flat, and crowded planet with our every move utterly dependent upon cheap and abundant energy. Without it, today's modern society would grind to a halt.

We all have benefited greatly from recent transformations in communication, transportation, and food production made possible by access to energy. Many of us now live longer and healthier lives. This modern lifestyle is so appealing that 6 billion people in developing countries around the world are trying to live it.

By 2050, the human footprint of energy development and the accompanying losses in ecosystem services and biodiversity will be massive—if we continue to extract that energy in the same way as we do now. Predictions show a possible 65 percent increase in energy demand by 2050. This means that regardless of the energy development path we choose—business as usual or renewables instead of fossil fuels—the resulting energy footprint will continue to vex us.

The harsh reality of our global energy future is the struggle to find a way to sustainably provide for a projected nine billion people by 2050. Just as the twentieth century is known for technological growth in support of the human experience, we of the twenty-first will be remembered for our ability to rapidly implement creative solutions to address resulting energy and climate challenges (figs. 0-1 and 0-2).

Now we are at a crossroads. We already have solid scientific findings that show the cumulative impacts from energy development to the environment. New tools to mitigate those impacts are available, but we haven't yet seen many of the benefits. And we won't fully realize those benefits until decision makers put these strategies into widespread use.

Pressure on Natural Environment

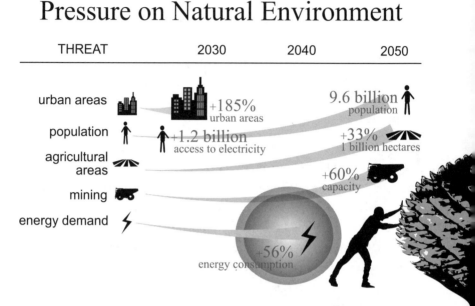

Figure 0-1. *Summary of global development pressures. Data adapted from Oakleaf et al., "A World at Risk" (see ch. 1, n. 6).*

Unfortunately, however, this issue has not received the attention it deserves. This void seems implausible when climate change, energy scarcity, and energy security have dominated front-page headlines for the past several years.

So why this book and why now? Simply put, because the implications of today's decisions will influence society for decades. The seeds of the next energy century must be sown now because of the long life spans of power plants, refineries, and other energy infrastructure. We must dive deeply into this issue if we are to reconcile our need for abundant and reliable energy and maintain the integrity of irreplaceable ecosystem services. This book is part of that deep dive.

We conceived this book soon after publication of David Naugle's 2011 volume *Energy Development and Wildlife Conservation in Western North America*. Joseph Kiesecker and other staff from The Nature Conservancy were major contributors to this development of a road map for energy development that safeguards people and nature.

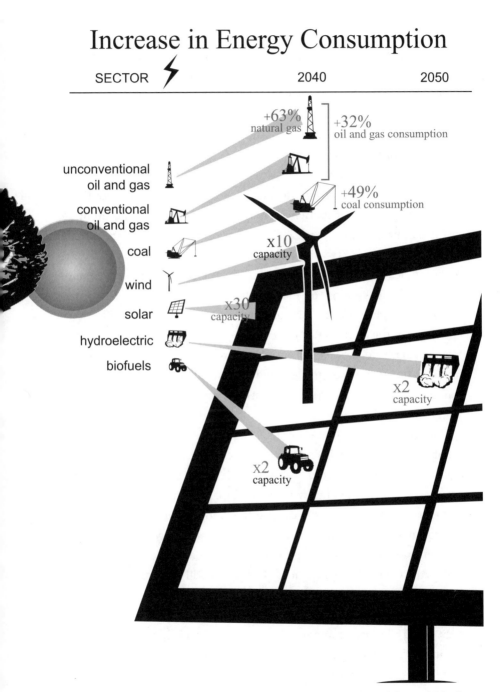

Increase in Energy Consumption

SECTOR ⚡	2040	2050

+63%
natural gas

+32%
oil and gas consumption

unconventional
oil and gas

+49%
coal consumption

conventional
oil and gas

coal

x10
capacity

wind

solar

x30
capacity

hydroelectric

biofuels

x2
capacity

x2
capacity

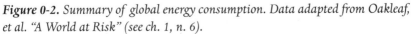

Figure 0-2. Summary of global energy consumption. Data adapted from Oakleaf, et al. "A World at Risk" (see ch. 1, n. 6).

Good ideas take time to mature—this new book is the one we wanted to write the first time. Five years ago we weren't prepared to go outside of western North America; now, however, we have multinational experience and contacts that enable us to revisit our original vision. Readers of the first book told us they wanted less about impacts of energy development and more on scalable and lasting solutions. We heard you—that's the subject of this book.

As the architect for The Nature Conservancy's signature program Development by Design and through his travels across five continents, Joseph Kiesecker has experienced firsthand the challenges posed by rapidly expanding development. He also sees opportunity for new frontiers in energy sustainability. His work in developed and developing countries makes it clear that we are ill prepared for the scale of impacts resulting from tomorrow's energy developments.

Dave Naugle is a University of Montana professor of wildlife and currently serves the Natural Resources Conservation Service as their national science adviser to the Sage Grouse Initiative. Since 2010, the Sage Grouse Initiative has brought unprecedented resources ($460 million) to the western United States to partner with private landowners and ranchers for wildlife conservation.

Our objective for this new book is to provide a road map for conservation that elected officials, industry representatives, natural resource managers, environmental groups, and the public can use to find the balance between securing our energy future and maintaining critical biodiversity and ecosystem services. We also see this book as a call to action for the environmental community to strategically shift away from its hardline opposition of energy development toward proactively guiding the right energy mix into the right places. The latter will be critical if we are to create more renewable energy sources and achieve ambitious emissions reduction targets. More than one billion people do not have access to electricity, and the environmental community, especially the large environmental NGOs based largely in developed countries, must not take stands that inhibit energy access in developing countries.

The book is divided into three sections. Part I lays out an overview of global energy patterns and global development scenarios to make a clear case for the issue as a global challenge. Part II highlights energy development case studies from around the world that show best practice approaches that could be replicated more widely. We selected case studies from seventeen countries spanning the range of geographies and energy sectors identified in part I. This includes focus on renewable and unconventional energy sources slated to increase dramatically in coming years. In part III we lay out a blueprint for what it will take to turn best practice into common practice.

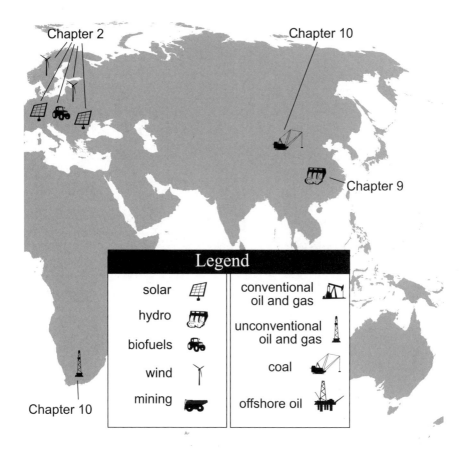

PART I

A Glimpse into Future Sprawl

By 2050, global energy consumption is expected to grow by over 65 percent and global electricity demand is projected to nearly double (fig. p1-1). This development will help fuel economic growth, improve quality of life, and lift people out of poverty—1.2 billion more people will gain access to electricity by 2030. This growth has clear benefits and seems destined to happen. What we don't know is how much, where, and what it will cost natural systems.

At the same time, climate change remains a serious concern—many people characterize it as the defining environmental challenge of the twenty-first century. Climate is central: glimpsing into our energy future, we see that we are going to use more of every type of energy to fuel a growing and more affluent world population. Efforts to reduce greenhouse gas emissions, the bulk of which are from energy-related activities, will strongly shape the future of energy development.

If the world wants to hit the carbon emission reduction targets we agreed to in the 2015 Paris Agreement, we'll have to shift from traditional carbon energy sources to a mix of renewables. We'll also then experience firsthand the trade-offs caused by this shift to renewables—namely, the heavier human footprint associated with renewable energies. Without careful planning we could trade one crisis—climate change—for another: land-use change and conflict (fig. p1-2).

To understand how and where future energy development might happen, we have to first understand more deeply our inventory of potential energy resources. No one is certain what energy mix will best serve both increasing demands and attempts to reduce CO_2 emissions, so we have to consider potential development scenarios or alternative development pathways. Part I of this book provides an industry perspective into

1

Shift in Global Energy Mix

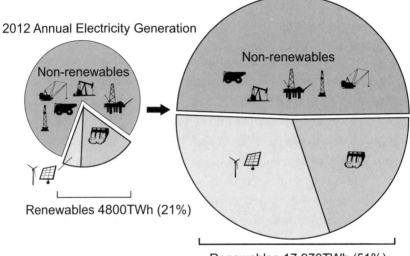

2040 Projected Electricity Generation

2012 Annual Electricity Generation

Non-renewables

Non-renewables

Renewables 4800TWh (21%)

Renewables 17,970TWh (51%)

Figure p1-1. Shift in global energy mix; data adapted from IEA, World Energy Outlook 2015 *(see ch. 1, n. 3).*

long-term trends in energy supply and demand, and into the resulting geopolitical shifts and social change that underlie our potential energy future. As part of the first-ever global analysis across all energy sectors, part I also looks at which continents and countries are likely to be most impacted by our energy decision making.

Readers of chapter 1 will uncover this key message about our ever-increasing energy footprint on earth: it's a lot! Our new estimates of the relative potential for energy expansion show that globally, 20 percent of remaining natural lands—about 20 million square kilometers, or an area larger than the size of Russia—are considered to be under high threat of future development. More than half the world's regions will face the risk of up to 50 percent conversion of their land surface to some sort of

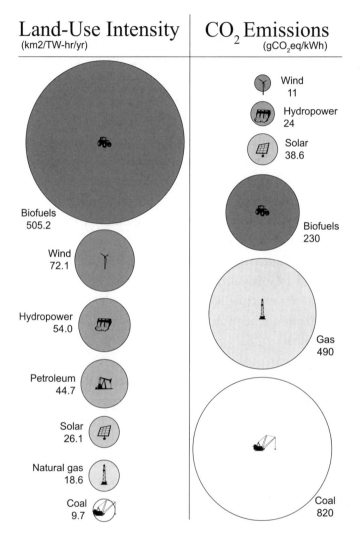

Land-Use Intensity
(km2/TW-hr/yr)

CO_2 Emissions
(gCO_2eq/kWh)

Wind
11

Hydropower
24

Solar
38.6

Biofuels
505.2

Wind
72.1

Hydropower
54.0

Petroleum
44.7

Solar
26.1

Natural gas
18.6

Coal
9.7

Biofuels
230

Gas
490

Coal
820

Figure p1-2. *Comparison of the energy footprints in terms of land required (km²/TW-hr/yr) and CO₂ emissions (gCO₂eq/kWh). Land footprint adapted from McDonald et al., "Energy Sprawl or Energy Efficiency" (see ch. 1, n. 4); CO₂ emissions data adapted from Steffen Schlömer, Thomas Bruckner, Lew Fulton, Edgar Hertwich, Alan McKinnon, Daniel Perczyk, Joyashree Roy, Roberto Schaeffer, Ralph Sims, Pete Smith, and Ryan Wiser, "Annex III: Technology-Specific Cost and Performance Parameters,"* in Climate Change 2014: Mitigation of Climate Change, *Working Group III Contribution to the Fifth Assessment Report of the Intergovernmental Panel on Climate Change, 2014.*

development. Africa and South America will likely undergo a doubling and tripling of natural lands impacted. Already-developed regions will not be immune to further development—people living in Europe, Central America, and Southeast Asia will live with a land-use intensity they have never before experienced. These impacts are likely to arise not from a single driver, but as a multitude of sectors acting in concert. Legal land protections are poorly equipped to counter this threat, with less than 5 percent of at-risk natural lands currently protected.

In chapter 2 we highlight two main alternative development scenarios: the first, in which proactive policy drivers dominate, forcing society to address energy issues in ways that will reduce impacts to nature; and the second, in which market drivers dominate, resulting in significant impacts to nature. If government plays a strong role and firm, far-reaching policy measures are introduced, we'll see more-compact cities and a transformed global transport network in the future. These new policies can help promote more renewable energy and lower-carbon energy sources (e.g., natural gas).

Alternatively, we could see a more volatile energy future. If energy demand surges, owing to strong economic growth, influence might be more widely distributed and government will take longer to agree on major decisions. Market forces rather than policies will shape this energy system; oil and coal remain a significant part of the energy mix and renewable energy grows more slowly. Without a conscious effort to alter the business-as-usual approach through regulations on energy use and production, we won't be able to avoid the impacts of continued energy development.

CHAPTER 1

Geography of Risk

James Oakleaf, Christina M. Kennedy,
Sharon Baruch-Mordo, and Joseph M. Kiesecker

Human populations have a tendency to sprawl. Just look out the plane window as you leave any airport and you will likely see a city blending with agricultural fields for miles. With the aid of satellite imagery, this pattern can be witnessed on a global scale, revealing that urban and agricultural areas now make up over 40 percent of the Earth's land surface.[1] Sprawl is also a critical issue for the energy sector. Energy sprawl is the product of the amount of energy produced and the land-use intensity of production. Production is the terawatt hours per year of energy and intensity is the square kilometers of habitat given over to that production.

Different types of energy vary widely in their footprint. Many of the renewable energy sectors that are desperately needed to combat climate change have very large spatial footprints. As human populations grow—expected to hit 9 billion by 2050[2]—and demand for energy soars in developing countries, something will have to give.[3] The larger energy footprint of renewables will inevitably lead to trade-offs with land-use conversion.[4] We will need to manage energy development in a way that can meet demands but also reduces the impacts on natural systems that support both human and wildlife populations.

Assessing Cumulative Risk

Proactively identifying lands at risk of conversion and strategically planning to mitigate future impacts is critical to achieve a sustainable balance
between development and conservation.[5] But striking this balance is only
possible if we first understand where and how future development may
occur. To do this we combined nine potential development threats to
identify where current natural lands are at future risk of conversion or
modification.[6] We aggregated spatial patterns of expected energy threats
from conventional and unconventional oil and gas, coal, solar, wind, hydropower, and biofuels, and merged these with nonenergy threats from
mining, urbanization, and agricultural expansion to produce a global cumulative development threat map (fig. 1-1a). Next, we determined where
high cumulative development threat overlapped with current natural
lands to identify habitats at future risk (fig. 1-1b). We summarized the
threat patterns by terrestrial biomes (the world's major ecological communities—e.g., grasslands or forests) and geopolitical regions to show
where proactive conservation planning and subsequent actions might be
beneficial. We combined energy and nonenergy footprints because natural areas without risk from one source of development may still be impacted by another, and proactively planning to mitigate for all potential
sources of habitat loss or fragmentation is the only way to maintain large,
intact landscapes for conservation and human well-being.

Natural Lands at Risk

New developments are likely to be dispersed across the globe, potentially affecting 20 percent of the Earth's remaining natural lands (fig.
1-1b).[7] Currently, 21 percent of all biomes have half their natural habitats converted, and 57 percent have more than a quarter converted. Based
on our study, future development could push half the world's biomes to
more than 50 percent converted, and all biomes could lose over 25 percent
of their natural lands (with the exception of boreal forests and tundra).

Although development risk is globally dispersed, we found that three
biomes could be disproportionally affected. These contain 66 percent of

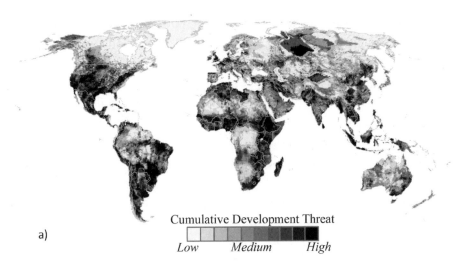

a)

Cumulative Development Threat

Low   Medium   High

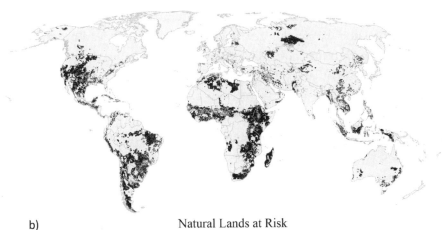

b)   Natural Lands at Risk

Figure 1-1. Future cumulative development threat and natural lands at risk.
(a) Global development threat map produced by equally weighting and combining
individual future sector development threat maps. (b) Natural lands at-risk
map produced by selecting natural lands that fell within the top 25 percent
of the cumulative threat scores. Adapted from "A World at Risk: Aggregating
Development Trends to Forecast Global Habitat Conversion."

delineated at-risk natural lands: tropical and subtropical grasslands, savannas, and shrublands (total area of 5.84 million square kilometers); deserts and xeric shrublands (total area of 3.7 million square kilometers); tropical and subtropical moist broadleaf forests (total area of 3.4 million square kilometers). Accounting for current and potential future development, three biomes could become predominantly human modified: tropical and subtropical dry broadleaf forests (83 percent); temperate broadleaf and mixed forests (72 percent); and mangroves (71 percent).

The geopolitical regions of Central America, Europe, and Southeast Asia face the most land conversion when factoring in future development threat. In contrast, Africa and South America, which are currently among the least modified regions, are predicted to have the highest amount of natural lands at risk (8.18 and 4.32 million square kilometers, respectively), potentially leading to future development doubling and tripling the amount of converted lands in South America and Africa, respectively.

Overall, only 5 percent of the at-risk natural lands have some form of legal protection that might prevent conversion. Existing land conversion combined with future potential habitat modifications highlight the need to act quickly in order to reduce impacts.

Energy Patterns

While our findings suggest future energy expansion across sectors is likely to be globally dispersed, there are sector-specific geographic patterns (fig. 1-2a–g).

Conventional Oil and Gas (COG). Worldwide, more than 100 countries are currently producing COG.[8] Since many of the top developed COG basins are also those with the highest volume of untapped resources, our analysis predicts that, to a large degree, future COG will mainly consist of intensification of the basins already producing oil or gas (fig. 1-2a). Offshore COG (not examined in our study) may be one exception, where many resources remain unexploited or are in early stages of development.[9]

Unconventional Oil and Gas (UOG). Recent technological advancements, including horizontal drilling in conjunction with hydraulic fracturing, have spurred a rapid increase in UOG production over the last decade.[10] However, many of the areas with potential resources remain unproven to a large degree owing to both the technological and economic feasibility to remove the resource. Despite this uncertainty, UOG development has helped the United States become a net energy exporter. We found that several other countries, including Argentina, Russia, Mexico, Australia, and China, are poised to be in a similar position (fig. 1-2b).

Coal. Five countries (United States, Russia, China, Australia, and India) contain over two-thirds of the world's coal reserves and produce 75 percent of the world's coal.[11] Although coal production and usage is expected to decline in response to global commitments to reduce CO_2 emissions, projections show coal as a staple of energy use for several decades.[12] We found that similar to conventional oil development, future coal development will consist largely of intensification in regions that are currently high producers, with those five top-producing countries likely to see the most coal expansion in the future (see fig. 1-2c).

Renewables. Renewable energy has the greatest potential for expansion and is estimated to surpass coal as the largest supplier of electricity by 2040.[13] However, siting will be a challenge given the land-use intensity of all renewable energy sectors.[14]

Biofuels. Most cropland expansion for biofuel production (or liquid fuels, mainly ethanol or biodiesel made from organic matter) will be located in tropical regions of the world.[15] Our assessment shows land conversion threats from biofuel development is highest in tropical South America and Africa and in select areas in Southeast Asia (e.g., Indonesia) (fig. 1-2d).

Hydropower. Hydropower produces over 16 percent of the world's electricity and is the dominant renewable energy source across the globe, currently employed in more than 150 countries.[16] We found that Southeast Asia, Africa, and South America are most threatened by future development—these countries happen to contain many of the longest free-flowing or undammed rivers (fig. 1-2e).

Solar. Two main types of technologies, concentrating solar power (CSP) and photovoltaic (PV), are used in utility-scaled solar power. Currently, solar power makes up less than 1 percent of global electricity demand but is the fastest growing renewable electricity sector.[17] Only 50 countries produce solar power on a commercial basis.[18] Our threat analysis shows that a large portion of the globe is suited for solar development, especially in much of Africa, the Middle East, India, Mexico, portions of Brazil and Chile, and the southwestern United States. (fig. 1-2f).

Wind. Wind power generation has more than doubled in the last four years and currently meets 4 percent of the global electricity demand.[19] More than 103 countries produce wind power on a commercial basis with high growth rates in the United States, China, and Eastern Europe.[20] These three regions also have high potential for future wind development, as do portions of South America, East Africa, New Zealand, and Japan (fig. 1-2e).

A Solution: Proactive Landscape-Scale Mitigation

Given the scale of potential future energy development, society needs to dramatically change the way we plan for, regulate, and mitigate these impacts. Legally protected areas might not be able to steer development away from sensitive areas.[21] Land-use planners must improve existing tools and create new approaches to address pending impacts. Environmental impact assessments (EIAs) are one of the chief tools currently used to mitigate impacts from energy development. EIAs, used with impact mitigation, are a systematic process to examine the environmental consequences of planned developments. These tools also emphasize prediction and prevention of environmental damage through the application of the mitigation hierarchy: avoid, minimize, restore, or offset.[22] However, mitigation tools are conventionally implemented through a narrow spatial lens, at a project or site level that often results in uncoordinated, piecemeal mitigation that fails to deliver conservation outcomes at relevant ecological scales.[23] Based on our assessment, we propose a shift in regulatory oversight with an eye toward regional-scale, cumulative impact

Development Threat Ranking

Low   Medium   High

a) Conventional Oil and Gas

b) Unconventional Oil and Gas

c) Coal

d) Biofuels

e) Hydro Power

f) Solar

g) Wind

Figure 1-2. Global energy development threat maps. We introduce a new threat map for hydropower following similar methods utilized in "A World at Risk: Aggregating Development Trends to Forecast Global Habitat Conversion," with data adapted from Christiane Zarfl, Alexander Lumsdon, Jürgen Berlekamp, Laura Tydecks, and Klement Tockner, "A Global Boom in Hydropower Dam Construction," Aquatic Sciences 77, no. 1 (January 2015): 161–70, doi: 10.1007/s00027-014-0377-0.

assessments and proactive mitigation planning that better accounts for future development threats from multiple sectors.

As development encroaches into more remote and previously undisturbed areas, corporations, governments, development banks, and civil society groups must collaborate to avoid and minimize future impacts on remaining habitats. In an effort to change future trajectories, we propose that environmental licensing, impact mitigation, and financing should target where development could impact significant proportions of natural areas. Mitigation requirements should include procedures for proactively evaluating the compatibility of proposed development with conservation goals to determine when impacts should be avoided and when development can proceed.[24] Given the expansive scale of expected impacts from a variety of sectors, developers will need to compensate for residual impacts through the use of biodiversity offsets.

Biodiversity offsets, also known as set-asides, compensatory habitat, or mitigation banks, can maintain or enhance environmental assets in situations where development is moving forward despite negative impacts.[25] Future analyses that look at natural areas at greatest risk to cumulative development threats should be performed at finer (landscape) scales—for example, as done by Saenz and others in "Development by Design in Colombia." (See also chapters 3, 4, 5, 6, and 9 in this book.) These should be used to guide the siting of offsets so that the most ecologically important and at-risk areas are secured.[26]

Acknowledgments

We thank Paul C. West and James S. Gerber (Institute on the Environment, University of Minnesota), Navin Ramankutty (Liu Institute for Global Issues, University of British Columbia), and Larissa Jarvis and Dany Plouffe (Land Use and Global Environment Research Group, McGill University) for providing technical assistance on agricultural and biofuel expansion threats; Bennett Holiday, Bob Barnes, and Bryan Woodman for helpful discussions; and all the data providers who publically provide their data, thus enabling research like ours. Funding for our analysis was provided by

The Nature Conservancy, Anne Ray Charitable Trust, and The Robertson Foundation.

Notes

1. Roger Hooke and José Martin-Duque, "Land Transformation by Humans: A Review," *GSA Today* 22, no. 12 (2012): 4–10, doi:10.1130/GSAT151A.1; United Nations Department of Economic and Social Affairs Population Division, *World Urbanization Prospects: The 2014 Revision*, 2014, https://esa.un.org/unpd/wup/Publications/Files/WUP2014 -Highlights.pdf; Karen Seto, Burak Güneralp, and Lucy Hutyra, "Global Forecasts of Urban Expansion to 2030 and Direct Impacts on Biodiversity and Carbon Pools," *Proceedings of the National Academy of Sciences of the United States of America* 109, no. 40 (2012): 16083–88, doi:10.1073/pnas.1211658109.

2. UN, "World Urbanization Prospects."

3. Seto et al., "Global Forecasts of Urban"; Patrick Gerland, Adrian E. Raftery, Hana Ševčíková, Nan Li, Danan Gu, Thomas Spoorenberg, Leontine Alkema, Bailey Fosdick, Jennifer Chunn, Nevena Lalic, Guiomar Bay, Thomas Buettner, Gerhard Heilig, John Wilmoth, "World Population Stabilization Unlikely This Century," *Science* 346, no. 6206 (2014): 234–37, doi:10.1126/science.1257469; Felix Creutzig, Giovanni Baiocchi, Robert Bierkandt, Peter-Paul Pichler, and Karen Seto, "Global Typology of Urban Energy Use and Potentials for an Urbanization Mitigation Wedge," *Proceedings of the National Academy of Sciences of the United States of America* 112, no. 20 (2015): 6283–88, doi:10.1073/pnas.1315545112; Intergovernmental Panel on Climate Change, *Climate Change 2014: Mitigation of Climate Change. Contribution of Working Group III to the Fifth Assessment Report of the Intergovernmental Panel on Climate Change*, 2014, http://www .ipcc.ch/report/ar5/wg3/; International Energy Agency, *World Energy Outlook 2015*, Nov. 10, 2015, http://www.worldenergyoutlook.org/weo2015/; Guangnan Chen, Gary Sandell, and Craig Baillie, "Improving Energy Efficiency in Agriculture," National Centre for Engineering in Agriculture, University of Southern Queensland, 2014.

4. Robert McDonald, Joseph Fargione, Joseph Kiesecker, William Miller, and Jimmie Powell, "Energy Sprawl or Energy Efficiency: Climate Policy Impacts on Natural Habitat for the United States of America," *PLoS ONE* 4 no. 8 (2009), doi:10.1371/journal .pone.0006802; Anna Trainor, Robert Mcdonald, and Joseph Fargione, "Energy Sprawl Is the Largest Driver of Land Use Change in United States," *PLoS One* 11 no. 9 (2016), doi:10.1371/journal.pone.0162269.

5. Joseph Kiesecker, Holly Copeland, Amy Pocewicz, and Bruce McKenney, "Development by Design: Blending Landscape-Level Planning with the Mitigation Hierarchy," *Frontiers in Ecology and the Environment* 8, no. 5 (2010): 261–66, doi:10.1890/090005.

6. James Oakleaf, Christina Kennedy, Sharon Baruch-Mordo, Paul West, James Gerber, Larissa Jarvis, and Joseph Kiesecker, "A World at Risk: Aggregating Development Trends to Forecast Global Habitat Conversion," *PLoS One* 10, no.10 (2015), doi:10.1371/journal.pone.0138334.

7. Ibid.

8. U.S. Energy Information Administration, *International Energy Statistics: Reserves of Oil and Natural Gas for 2012*, Nov. 2015, http://www.eia.gov.

9. U.S. Geological Survey, *Supporting Data for the U.S. Geological Survey 2012 World Assessment of Undiscovered Oil and Gas Resources*, U.S. Geological Survey World Conventional Resources Assessment Team, Nov. 1, 2013, http://pubs.usgs.gov/dds/dds-069/dds-069-ff/.

10. U.S. Energy Information Administration, *Technically Recoverable Shale Oil and Shale Gas Resources: An Assessment of 137 Shale Formations in 41 Countries Outside the United States*," Sept. 24, 2015, http://www.eia.gov/analysis/studies/worldshalegas/pdf/fullreport.pdf.

11. U.S. Energy Information Administration, *International Energy Statistics: Reserves of Oil and Natural Gas for 2012*, Nov. 2015. http://www.eia.gov.

12. IEA, "World Energy Outlook 2015."

13. Ibid.

14. McDonald et al., "Energy Sprawl"; Trainor et al., "Energy Sprawl."

15. "Coal Reserves, International Energy Statistics," U.S. Energy Information Administration, accessed January 15, 2014, http://www.eia.gov/cfapps/ipdbproject/IEDIndex3.cfm?tid=1&pid=7&aid=6.

16. United Nations, *Energy Statistics Yearbook 2013*, Feb. 2016.

17. Center for Climate and Energy Solutions, Solar Power factsheet, 2012, http://www.c2es.org/technology/factsheet/solar.

18. United Nations, *Total Solar Electricity Production by Country*, 2012, http://data.un.org/Data.aspx?q=solar&d=EDATA&f=cmID%3aES; U.S. Energy Information Administration, *Solar, Tide and Wave Electricty Generation by Country*, 2012, http://www.eia.gov/cfapps/ipdbproject/iedindex3.cfm?tid=6&pid=36&aid=12&cid=regions&syid=2010&eyid=2010&unit=BKWH.

19. World Wind Energy Association, *World Wind Energy—Statistics*, 2014, http://www.wwindea.org/home/index.php?option=com_content&task=blogcategory&id=21&Itemid=43.

20. Trainor et al., "Energy Sprawl."

21. Stuart Chape, Jeremy Harrison, Mark Spalding, and Igor Lysenko, "Measuring the Extent and Effectiveness of Protected Areas as an Indicator for Meeting Global Biodiversity Targets," *Philosophical Transactions of the Royal Society B* 360, no. 1454 (2005): 443–55, doi:10.1098/rstb.2004.1592; Lucas Joppa and Alexander Pfaff, "Reassessing the Forest Impacts of Protection: The Challenge of Nonrandom Location and a Corrective Method," *Annals of the New York Academy of Sciences* 1185 (2010): 135–49, doi:10.1111/j.1749-6632.2009.05162.x.

22. Bruce McKenney and Joseph Kiesecker, "Policy Development for Biodiversity Offsets: A Review of Offset Frameworks, *Enviromental Mangement* 45, no. 1 (2009): 165–67, doi:10.1007/s00267-009-9396-3.

23. Jessica Wilkinson, James McElfish Jr., Rebecca Kihslinger, Robert Bendick, and Bruce McKenney, "The Next Generation of Mitigation: Linking Current and Future Mitigation Programs with State Wildlife Action Plans and Other State and Regional Plans," Environmental Law Institute white paper, Aug. 2009.

24. McDonald et al., "Energy Sprawl"; Shirley Saenz, Tomas Walschburger, Juan Carlos González, Jorge León, Bruce McKenney, and Joseph Kiesecker, "Development by Design in Colombia: Making Mitigation Decisions Consistent with Conservation Outcomes, *PLoS One* 8, no. 12 (2013), doi:10.1371/journal.pone.0081831.

25. McKenney and Kiesecker, "Policy Development."

26. Joseph Kiesecker, Holly Copeland, Amy Pocewicz, Nate Nibbelink, Bruce McKenney, John Dahlke, Matt Holloran, and Dan Stroud, "A Framework for Implementing Biodiversity Offsets: Selecting Sites and Determining Scale," *Bioscience* 59, no. 1 (2009): 77–84, doi:10.1525/bio.2009.59.1.11.

CHAPTER 2

Challenges of a Green Future

Gert Jan Kramer

In the book *Nature's Economy*, the intellectual historian and writer Donald Worster describes how humans' view of nature has been bookended by two different intellectual traditions through the centuries: Arcadianism and imperialism.[1] The idea of Arcadia is inspired by humans' desire to live in harmony with nature, while imperialism represents the equally human urge to dominate it. Most of us are torn between the two, dreaming about and striving for harmony with nature, yet in our actions we are utter imperialists.

In this chapter we explore the challenge of building a sustainable energy system. As we work toward a sustainable and mostly renewable energy system, we're perhaps guided by those Arcadian values, but the reality of such a system is steeped in the imperialism that has reshaped the Earth over the past millennia. A green future won't necessarily be one in perfect harmony with nature. Industrialization of the landscape is perhaps inevitable in the fight against a greater evil: a landscape completely changed for the worse by climate change.

Building a mostly renewable energy system demands concrete planning (and actual concrete!). Much of the talk about the energy transition overlooks the land requirements needed to build the solar panels and wind turbines and to grow the biofuels that will produce gigawatts and reduce CO_2. In this chapter we provide first-order estimates of the acreage that will be covered by solar panels, dotted with wind turbines, and inundated for hydro reservoirs, and the vast land claim associated with the sustainable production of bioenergy.

Plans on the Map

Land-use estimates for energy must start with projections of future energy use. For this we use the Shell New Lens Scenarios,[2] along with more recent work by Shell that explores the makeup of an energy system with (net) zero greenhouse gas emissions.[3]

Once we have an estimate of the future energy requirements, we can "put the plan on the map," to use a phrase coined by David MacKay, who did this for the United Kingdom.[4] All that is required are estimates of how much the various renewable energy sources produce each year per square kilometer. Such estimates are necessarily indicative, as the numbers will vary from place to place (as for instance the difference in photovoltaic yield between sunny and not-so-sunny locations) and between different authors. We base ourselves here on work by Vaclav Smil, augmented by our own earlier estimates.[5]

With these inputs, a first estimate of the impact of the energy transition can be done and—for Europe—is shown in figure 2-1.

In addition to a physical footprint, a renewable energy system will require greater integration across political boundaries given the variation in resource potential and seasonal variability of renewables. The European Climate Foundation's Roadmap 2050 illustrates this well.[6] The report shows that the key requirements for an efficient and effective renewables-dominated European energy system are a regional differentiation of renewable energy production according to the local resource (in particular, wind in the north and solar in the south) and a strong physical integration of the energy grid across Europe to deal with the momentary and seasonal variability of renewables.

For the United States, a more near-term look at land-use impact of new energy deployment shows that nations have a choice as to how they develop their energy system.[7] In the paper "Energy Sprawl or Energy Efficiency," the authors point out that in the absence of a strong focus on

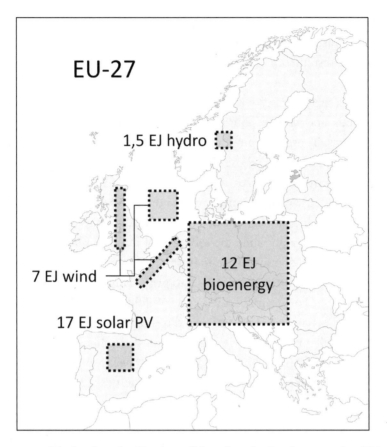

Figure 2-1. *The land-use implications of deep decarbonization scenarios. This one is based on Shell's Oceans scenario, circa 2100. The total energy use is circa 66 EJ/year, of which two thirds comes from renewables (from* New Lens Scenarios*). The energy production density estimates are based on* Energy in Nature and Society: General Energetics of Complex Systems.

energy efficiency, energy policy targets in the United States would—already by 2030—impact 200,000 square kilometers of land in the lower forty-eight states.

As the energy transition unfolds over the course of this century, the landscape will also undergo a transition. Humans will likely resist this transformation of the landscapes that we know, love, and cherish. But we need to accept some changes in order to stave off the far more threatening and devastating modification of the landscape (and of nature itself) that would result from climate change if we do *not* overhaul the energy system.

This presents society with a dilemma. We are emotionally attached to the landscape we have, but we are equally attached to our consumption patterns. These consumption patterns are underwritten by copious amounts of fossil fuels, whose land footprint is small relative to renewables. People might be surprised to hear this, but the average shale oil well has an energy production more than ten times what a modern wind turbine can produce. The global fossil fuel infrastructure would fit within the land area of Qatar, while the footprint of a future renewable energy system will have a continent-size footprint (fig. 2-2).

Society has hardly begun to come to grips with this aspect of the energy transition. For example, the World Wildlife Fund highlights major lifestyle and behavioral changes needed to reach a renewable energy future.[8] They call out two lifestyle changes critical to achieving this goal: the reduction of both meat consumption and air travel. Neither looks like an easy sell to an ever more affluent world community.

Biofuels and bioenergy stand out as the most prominent land-use challenge for a sustainable energy system. But photovoltaics and wind are not without their challenges—as anyone who has seen the acrimonious fights over wind turbines in their municipality can attest. One way to delve into this complex dilemma is through energy scenarios. Scenarios can demonstrate for us different pathways toward a low-carbon energy system.

Energy Scenarios

One approach to assessing the future prospects of different energy technologies and how they might reshape the energy system is to ask whether the energy system will be rebuilt top-down or bottom-up. Are governments pulling the strings, driven by supranational climate agreements and national energy agendas? Or are consumers and producers of energy ultimately in charge, through their purchase and investment decisions?

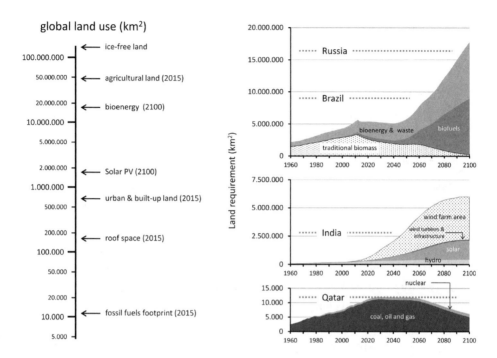

global land use (km²)

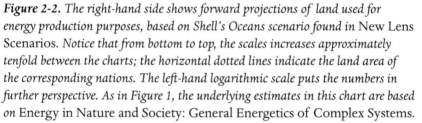

Figure 2-2. *The right-hand side shows forward projections of land used for energy production purposes, based on Shell's Oceans scenario found in* New Lens Scenarios. *Notice that from bottom to top, the scales increases approximately tenfold between the charts; the horizontal dotted lines indicate the land area of the corresponding nations. The left-hand logarithmic scale puts the numbers in further perspective. As in Figure 1, the underlying estimates in this chart are based on* Energy in Nature and Society: General Energetics of Complex Systems.

In the latest set of Shell energy scenarios, the top-down narrative is called Mountains, and the bottom-up story is named Oceans.[9] In Mountains, governments and powerful stakeholders, aware of the need to act decisively on climate change, try to change the energy system to optimize costs across the full energy system. Ideally this means investments phased in time so as to minimize early write-off of existing infrastructure and with the pace of new technology deployment set by their commerciality. As always, reality falls short of the aspirations, but in the Mountains scenario, carbon price is put in place early, allowing private enterprise to build out new energies while avoiding as much as possible disruption of the existing system.

The energy choices that fit this model are

- A rapid switch from coal to gas
- The concurrent deployment of carbon capture and storage and nuclear alongside renewables
- The development of hydrogen as a carbon-neutral energy carrier

These are set against the backdrop of a world in which cities follow a compact development model and moderate the energy service demand to keep the energy requirement relatively low.[10]

The Oceans scenario, by contrast, paints a world in which the initiatives of individuals and companies, encouraged by patchwork of (local) government initiatives, are the main drivers of change. Each does what it can, often without a clear combined plan at the (inter) national level. This is a world in which global agreements are elusive or ineffective, but the local actors are driven by a care for the environment, concern about climate change, and a desire for energy independence in a fragmented world. The outcome is not necessarily efficient, but it does deliver a strong growth of renewables.

Photovoltaics are the most obvious winner, as solar is appealing and accessible to everyone, everywhere. Another winner is onshore wind. As wind systems grow, their intermittent power output will increasingly burden the power system, but this is managed by ad hoc storage and demand-management measures. Biofuels are a winner, too, albeit with marked regional variation: for some countries, especially those with low to medium population densities, biofuels are a renewable, local, and secure fuel. Rapid renewables deployment is the positive side of bottom-up. The downside is a failure to lower the carbon footprint of fossil fuels, which remain a significant part of the mix for decades to come.

Neither the switch from coal to gas, nor the development of carbon capture and storage are priorities in the Oceans scenario. They are seen as too expensive (coal-to-gas) or pointless and unaffordable (carbon capture and storage) in the absence of a transnational climate and carbon pricing agreement. As the story unfolds through time, the world eventually turns to them—out of necessity.

Maxing Out the Energy Mix

More recent work of Shell's scenario group shows how difficult it is to replace hydrocarbons across the full spectrum of energy services. Electrification of personal transport and of the home is possible and might eventually be nearly universal. But heavy industry, long-distance freight transport, and aviation will continue to rely on hydrocarbons for lack of practical alternatives. Without a massive and global change in lifestyle and energy service demands, hydrocarbons could easily be a third or more of the world's primary energy—350 out of 1,000 exajoules (EJ).[11]

This analysis predicts that at most approximately 200 EJ of primary biomass energy will be available for energy purposes—not quite half of what is likely to be required.[12] This leaves the world with no other choice than to develop carbon capture and storage technology—unless we can make up for the shortfall of bio-based hydrocarbons by making them artificially. This inspires the dream of artificial photosynthesis,[13] which is unfortunately not yet developed enough to see it as a get-out-of-jail-free card.

Both the work on net-zero emissions and the long-term developments in Mountains and Oceans show a diverse energy mix toward the end of the century. In order to deliver 1,000 EJ of primary energy for the full range of energy services at net-zero emissions, all available energy resources must be deployed to their maximum acceptable potential. These limits include for solar and wind, as much as can be accommodated in the system; for hydropower and geothermal energy, as much as is available; for bioenergy, as much as sustainable land use and the requirements for food, feed, and fiber allow; for nuclear, as much as the national governments choose to champion it; and finally, for fossil fuels, where they are irreplaceable, and with carbon capture and storage to mitigate and net out emissions to zero.

Land Claims and Governance

The estimated consequences for land use are mindboggling. In Shell's analysis, about half of the 1,000 EJ primary energy can come from non-biomass renewables, with solar, wind, and hydro as the major sources. The land requirements and the development over time, according to the Oceans scenario, are shown in figure 2-2. The analysis also predicts about half the 1,000 EJ of primary energy can come from nonbiomass renewables, and biomass is used to its 200 EJ maximum. This leaves about a third to be supplied from nuclear (about 8 percent, or 80 EJ, well over two times today's nuclear energy production) and approximately a quarter from fossil energy—25 percent, or 250 EJ, just about half of what the world uses today (in an energy system that by 2100 will be twice as large as today's).

Scenarios paint different pictures of the future, and so it's no surprise that plots of emissions and energy land-use requirements show different trajectories over time (fig. 2-3).

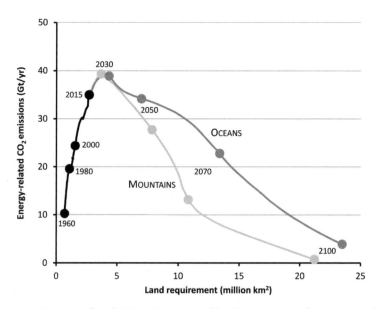

Figure 2-3. *Energy-related CO$_2$ emissions and land requirement for energy as they have developed over time, and as they might develop into the future, according to the Shell scenarios Mountains and Oceans.*

The divergence in the midterm creates a conundrum for the management of energy and land, especially for the Oceans scenario: the weak global governance in this scenario must somehow be paired with strong—or at least effective—land-use management, lest the development of renewables runs afoul. Poorly planned large-scale photovoltaic and wind projects will be met with resistance, and large-scale bioenergy production may even be counterproductive when land-use change and agricultural practices are poorly managed.

A Simple Conclusion

We made a first, simple estimate of how low-carbon energy scenarios plot out on the map (figs. 2-2 and 2-3). The main point is simply this: it's a lot of land! The fact that it is a lot clearly underscores the need for planning that will guide site selection and will mitigate impacts of energy development that is the focus of part II of this book.

Acknowledgments

I conceived this paper while working for Shell and am indebted to my Shell colleagues, notably from the Scenarios team and the Future Energy Technologies team, for illuminating discussions. At the time of writing I moved to Utrecht University, where I hold the chair of Sustainable Energy Supply Systems at the Copernicus Institute of Sustainable Development.

Notes

1. Donald Worster, *Nature's Economy—A History of Ecological Ideas,* 2nd ed. (Cambridge, England: Cambridge University Press, 1994).
2. Shell, *New Lens Scenarios* (Shell, 2013).
3. Shell, *A Better Life with a Healthy Planet* (Shell, 2016).
4. David MacKay, *Sustainable Energy—Without the Hot Air* (Cambridge, England: UIT, 2009).
5. Vaclav Smil, *Energy in Nature and Society: General Energetics of Complex Systems* (Cambridge, MA: MIT Press, 2007); Jose Bravo and Gert Jan Kramer, "The Energy Density Conundrum," in *The Colours of Energy: Essays on the Future of Energy in Society*, edited by Gert Jan Kramer and Bram Vermeer (Shell, 2015).
6. European Climate Foundation, "Roadmap 2050: A Practical Guide to a Prosperous, Low-Carbon Europe" (ECF, 2010).
7. McDonald et al., "Energy Sprawl" (see chap. 1, n. 4).

8. World Wildlife Fund, "The Energy Report—100% Renewable Energy by 2050" (WWF, 2011).

9. Shell, *New Lens Scenarios*.

10. Shell, *New Lenses on Future Cities—A New Lens Scenario Supplement* (Shell, 2014).

11. Shell. *A Better Life with a Healthy Planet* (Shell, 2016).

12. Yvonne Deng, Michèle Koper, Martin Haigh, and Veronika Dornburg, "Country-Level Assessment of Long-Term Global Bioenergy Potential," *Biomass and Bioenergy* 74 (2015): 253–67, doi:10.1016/j.biombioe.2014.12.003.

13. Thomas Faunce, Stenbjorn Styring, Michael Wasielewski, Gary Brudvig, Bill Rutherford, Johannes Messinger, Adam Lee, Craig Hill, Huub deGroot, Marc Fontecave, D. MacFarlane, Ben Hankamer, Daniel Nocera, David Thiede, Holger Dau, Warwick Hillier, Lianzhou Wang, and Rose Amal, "Artificial Photosynthesis as a Frontier Technology for Energy Sustainability," *Energy and Environmental Science* 6 (2013): 1074–76, doi:10.1039/C3EE40534F.

PART II

Solutions for Reducing Energy Sprawl

Modern society is completely dependent on continuous flows of fuels and electricity from the combustion of fossil sources. Without sufficient energy to heat and light our homes, run our businesses, power manufacturing plants, and stoke our cars and planes, our world would come to a standstill.

Our century, more than any other time in history, is marked by decisive departures from long-lasting patterns, all driven by dramatic increases in energy usage. These range from revolutions in food production (mechanization, fertilizers, and pesticides) and transportation (cars, air travel) to even more rapid transformation of communication (radio, TV, satellites, and the Internet). Practically all pre-twentieth-century technological advances had gradual changes on society, but exploiting fossil fuel energy has rapidly and dramatically altered the trajectory of civilization. People have benefited greatly from this transformation—many of us live longer and have a higher standard of living.

Given the human benefits of development, massive energy expansion and production are inevitable. Undoubtedly, reduced consumption and increased energy efficiency will need to be part of the solution, but after reading part I of this book, readers will arrive at two sweeping conclusions: The first is that society must quickly recognize and subsequently embrace the inevitability of accelerated development as the new normal. Society remains largely uninformed about the consequences of this escalating development. Even more troublesome is the fact that many environmental groups in developed countries still champion the romantic yet unrealistic notion that everything can be saved for conservation. The second realization is that we need a better approach to planning energy developments to conserve biodiversity, clean water, and other ecosystem services. This is also is a plea for the conservation community to engage with industry to manage the future footprint, regardless of the energy pathway the world takes.

Without a better approach to energy planning, the story will remain the same: A company quietly acquires land rights, plans its individual project, and afterward complies with biodiversity laws and statutes in a piecemeal fashion. The ensuing lawsuits slow development, deplete conservation and industry resources that could be put to better use, and most times do not safeguard the natural resources valued by society. This piecemeal approach could unfortunately slow the proliferation of renewable energy—which is key to combat climate change. Environmental groups and industry alike are tired of this approach, and society is slowly waking up to ask the vexing yet crucial question: how can we expedite more responsible development and still maintain functional and connected ecosystems and their associated services and wildlife values?

The answer to this question is regional conservation planning undertaken across geographies and multiple energy sectors. Regional planning changes the old method of development by inverting the timing and scale at which biodiversity and other societal values are considered by industry. In this planning scheme, industry considers the potential impacts of multiple projects before they have been individually planned and are moving toward implementation. Regional planning looks at the bigger geographic picture, with individual projects planned collectively to avoid, reduce, and then mitigate undue impacts. Regional rather than fragmented planning results in more effective conservation outcomes, reduces regulatory hurdles for industry, and provides cost savings to conservation and industry.

Part I presented an overview of the spatial overlap between competing energy and biodiversity values and showed us a scary vision of the ghost of the energy future. Like Scrooge in *A Christmas Carol*, we can still alter our ways to create a better future. Part II provides options to resolve this conflict through creative and viable solutions that improve the future energy footprint.

Part II advances next-generation planning concepts using best-practice examples from seventeen countries, spanning the range of social and political situations and diversity of affected biological systems identified as priorities in part I. This framework blends a landscape vision with the mitigation hierarchy for a sustainable future. The approaches outlined in the following chapters all first avoid or reduce impacts on landscapes with irreplaceable wildlife or ecosystem service values, then minimize and restore impacts on-site using the best available technology, and, finally, offset residual impacts. Scenario modeling is an important component of regional planning because it allows decision makers to proactively examine the consequences of development. In chapters 4, 7, 8, and 9, we examine how these and other tools help decision makers envision consequences of their development choices and proactively prevent problems before they occur. We also use examples in chapters 7 through 9 to show readers how biodiversity offsets can mobilize the unprecedented level of available funding for conservation in a collective energy future.

While fossil fuels will likely still play a role in helping meet energy goals, development will increasingly expand using newer unconventional and renewable technologies. Chapters 3, 6, and 7 focus on how to improve siting and mitigation involved with traditional fossil fuel extraction. Chapters 4, 5, 8, and 9 focus on unconventional and renewable energy development. Most of this development will be on land, but some dramatic growth is expected in marine and freshwater environments. Chapter 6 examines how advanced conservation planning can promote incorporation of appropriate siting principles for offshore energy development. Chapter 9 examines how scenario planning can guide the placement of dams to preserve biodiversity and ecosystems service values. Chapter 9 also lays out how improved design and operation of dams can promote natural hydrologic flow regimes of the river systems they impact.

Even after reading about the better conservation outcomes possible with regional conservation planning, readers will finish part II with the somber realization that society still has a long way to go. For example, trade-offs with renewables remain evident in part II chapters: the cost of reducing emissions to combat climate change is an increase in the overall

Current Approach

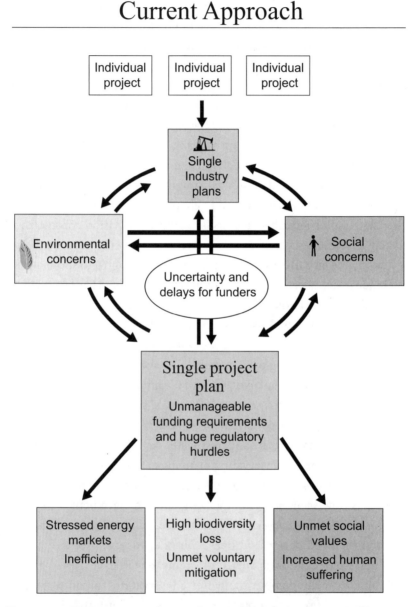

Figure p2-1. Business-as-usual energy licensing and planning process. This piecemeal process depletes conservation and industry resources, results in lawsuits, does not maintain natural resources valued by society and ultimately slows the proliferation of renewable energy.

Desired Landscape Planning

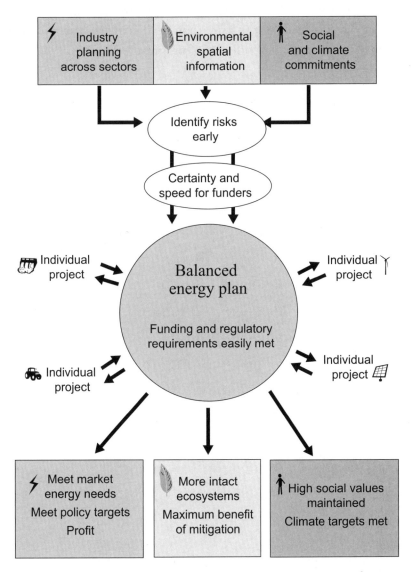

Figure p2-2. *Proactive and comprehensive regional energy planning. Before development investments are made, stakeholders consider the potential impacts of multiple projects that will drive more effective conservation outcomes, reduce regulatory hurdles, and result in cost savings to conservation and developers alike.*

human footprint on a per kilowatt basis. Obvious, too, is that despite documented benefits to conservation, most regional planning is still being done sector by sector. We're providing examples of the large-scale sector planning that has been done to date, knowing that in five years this might not look like much; there is still a lot of progress to be made.

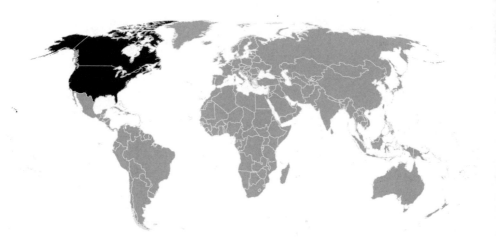

CHAPTER 3

Energy Sprawl and Wildlife Conservation

Mark Hebblewhite

North American wildlife conservation in the twentieth century often progressed under the implicit assumption that full recovery and conservation of species was possible. In the twenty-first century, society's increased energy demands are a wake-up call to facilitate conservation in the face of massive new development objectives. These two case studies from the sagebrush and boreal ecosystems of North America illustrate the benefits of proactive, rather than retrospective, conservation planning using triage frameworks.

The first case study, about federally threatened boreal woodland caribou (*Rangifer tarandus caribou*) and oil and gas development in Alberta, Canada, illustrates the consequences of decades of "save it all" mentality combined with the assumption that all industrial impacts can be mitigated. I review imminent extirpation for approximately half of Alberta's caribou populations, and huge impacts to biodiversity and ecosystem function that are only now being salvaged in a retrospective conservation strategy that is awkwardly out of step with regulatory policy. The second case study, that of the greater sage-grouse (*Centrocercus urophasianus*), illustrates the benefits of proactive conservation planning for the highest-quality sage-grouse habitat that prevents the need for an endangered species listing.

Boreal Woodland Caribou in Alberta

Alberta's oil and gas energy industry is key to the economies of Canada, the United States, and indeed the globe. Alberta is the engine of Canada's natural resource–based economy, producing a quarter of Canada's GDP by exporting more than 90 percent of its energy production. This province alone is the top supplier of foreign oil to the United States. Oil and gas produced there is a combination of conventional oil, natural gas, and the tar (or oil) sands. With over 173 billion barrels of tar sands, Canada is second only to Saudi Arabia in global oil reserves. The tar sands are the least efficient type of conventional oil, requiring double the energy per unit of energy extracted, and also the single biggest point source of CO_2 in North America.[1] James Hansen, former NASA chief scientist, called development of the tar sands "game over" for climate change.[2] The stakes of Alberta's energy industry are enormous for the globe.

In the northern half of the province, Alberta's vast energy infrastructure overlays the boreal forest, home to high biodiversity[3] and one of the largest remaining carbon stores of the planet.[4] As perhaps its leading flagship and umbrella species,[5] the boreal woodland caribou requires huge tracts of old-growth boreal forest to avoid predation and to survive on its specialized diet of slow-growing lichen. Most (90 percent) of Alberta's caribou range is on land leased by the government for development. In Alberta, the energy industry has created thousands of kilometers of access roads, pipelines, seismic exploration lines, and tens of thousands of well sites (fig. 3-1). Enabled by cheap access, the forest industry has further created a checkerboard of clear-cuts.

Both the roads and the clear-cuts directly reduce the amount of old forests upon which caribou depend. But humans' indirect impacts take the greatest toll on the caribou. Enhanced populations of early-seral species like moose (*Alces alces*) and white-tailed deer (*Odocoileus virginianus*) buoy numbers of carnivores like wolves (*Canis lupus*), which then drive down caribou populations through the mechanism of apparent competition across Canada (fig. 3-1).

Figure 3-1. *Conceptual figure showing effects of increasing habitat destruction and fragmentation on boreal woodland caribou in western Canada's boreal forest. Habitat destruction caused by energy development and forestry leads to increased primary prey and predator abundance, which in turn causes apparent competition-induced declines. Since 2006, Alberta has killed more than 1,000 wolves through aerial shooting and poison in the only effective mitigation strategy used to date to prevent continued caribou endangerment. More than 90 percent of the oil and gas produced in Alberta is exported to United States refineries to meet energy demand there and for further export. Credit: EHIllustration.com.*

In 2002, a new Species-at-Risk Act required that the federal and provincial governments respond to declining caribou numbers. They moved slowly, however: it took ten years for Canada to develop a recovery plan for the threatened boreal woodland caribou populations across the country.[6] The goal of federal/provincial recovery plans is central to this case study: to recover each and every single boreal caribou population in Alberta and the country. Based on the most comprehensive scientific effort to define critical habitat for an endangered species in Canada, the recovery plan determined that in Alberta, eleven of thirteen caribou populations are declining rapidly, by 50 percent every eight years over the last two decades.[7]

On average, 72 percent of all caribou range in Alberta is within 500 meters of human disturbance, exceeding a critical development threshold as stated in the federal recovery plan.[8] To date, in Alberta, there have been dozens of attempted mitigation strategies that tried to reduce impacts of ongoing development such as drilling restrictions in time and space, clustered development, temporary roads, and others. But only large-scale wolf control in the most developed caribou range, the Little Smoky herd, where more than 1,000 wolves have been aerially shot and poisoned since 2006, has been effective in halting caribou declines.[9] This desperate measure only served to buy time, during which 227 new gas wells[10] and thousands of hectares of new cutblocks were created, failing itself to recover caribou.

Only recently has the feasibility of the "save it all strategy" been questioned. Given the rapid declines in caribou numbers and the heavy industrialization of their habitat, it would take aggressive long-term habitat restoration and formal protection to reverse declines, probably combined with short-term predator control.[11] Based on the net 2010 value of just the energy leases (not profits), the protection strategy would cost $162 billion (in Canadian dollars)[12] to protect all Alberta caribou ranges. These costs however, do not include profit, which, for just the Little Smoky herd were estimated to run to $275 billion (2016 Canadian dollars)[13] for thirty years. Costs are highest for caribou ranges affected primarily by tar sands development (table 3-1). However, spatial prioritization schemes illustrate that by preferentially investing in the lowest-cost caribou strategies,[14] some caribou conservation could be achieved. Thus, the government faces a dilemma: federal and provincial Species-at-Risk Act recovery plans commit to recovering every single caribou population, but there is no economically feasible way they can.

A shift from a neoconservative to a socialist New Democrat governmental regime in 2016 removed many roadblocks to innovative thinking about caribou recovery. The new government released a pioneering independent mediators report on caribou conservation in Alberta.[15] This report ultimately took a strategic approach to caribou conservation, directing investments to the least developed herds in the northern parts of

Table 3-1. Status of at-risk caribou populations in Alberta, Canada[a]

Population	Caribou population growth rate[b]	Caribou N[c]	Primary energy development	Total disturbance (%)[d]	New oil and gas wells[e]	Protection[f]	Premier's report recovery actions[g]
YATE	1.001	300	Oil and gas	61	5	N/A	Protected area
ALP	0.939	150	Oil and gas	56	24	13	Wolf control, restoration, forestry phase-out
BIS	0.842	300	Oil and gas	71	75	739	Protected area
CM	0.92	400–500	Oil and gas	57	N/A	1165	Protected area
RRPC	0.892	325	Oil and gas	N/A	34	514	New energy lease deferral
LS	0.97	80	Oil and gas	95	213	656	Wolf control, restoration, forestry phase-out
RE	0.884	250–300	Oil and gas	62	234	7334	Protected area
CHIN	0.887	250–300	Oil and gas	76	43	8958	Protected area
SL	N/A	75	Oil and gas	80	1	2274	Business as usual
WSAR	0.933	300–400	Tar sands	69	1489	44,663	Business as usual
ESAR	0.916	150–200	Tar sands	81	1435	50,117	Business as usual
CL	0.855	100–150	Tar sands	85	3847	15,546	Business as usual
RICH	0.984	Unknown	Tar sands	82	717	30,905	Business as usual

[a] Retrospective triage in Alberta's caribou populations, showing each caribou population name and size; primary form of energy development; percentage of the range disturbed by human development; number of new wells drilled in the three years following release of the 2012 recovery plan; costs from Schneider et al. (2010) of protection (buying out energy leases); and what the recently released Alberta premier's report adopts for caribou recovery actions—the semblance of a quiet strategic plan.

[b] Caribou population growth rate from 3 to 18 years during 1994–2012, duration varying by herd (D. Hervieux, M. Hebblewhite, N. J. DeCesare, M. Russell, K. Smith, S. Robertson, and S. Boutin, "Widespread Declines in Woodland Caribou," *Canadian Journal of Zoology* 91, (2013): 872–82, doi:10.1139/cjz-2013-0123).

[c] Approximate caribou population size in 2006 from ASRDaAC Association, "Status of the Woodland Caribou (*Rangifer tarandus caribou*) in Alberta: Update 2010" (Edmonton. AB, 2010).

[d] Total disturbance of each caribou range within 500 meters of human development and burns as determined by the federal recovery strategy, from E. Canada, "Recovery Strategy for the Woodland Caribou (*Rangifer tarandus caribou*), boreal population, in Canada" (Ottawa: Environment Canada, 2012).

[e] Source: IHS dataset courtesy of B. W. Allred (B. W. Allred et al., *Science* 348 (April 24, 2015): 401.

[f] From R. R. Schneider, G. Hauer, W. L. Adamowicz, and S. Boutin, *Biological Conservation* 143 (July 2010), 1603.

[g] As provided in the Alberta mediators report, "Setting Alberta on the Path to Caribou Recovery."

the province, while remaining noticeably silent for the most expensive herds impacted by development. Key to these investments is the creation of 18,000 square kilometers of wildland protected areas, a phasing out of forestry in two of the most rapidly declining caribou ranges, and a $40 million Canadian fund in 2016 dollars for restoration activities that would favor caribou.

Unfortunately, such a triage approach is presently illegal under the "save it all" regulatory structure. This retrospective case study illustrates the challenges to strategic conservation planning in which recovery goals naively aim to save everything. Eventually this silent triage approach will be tested, and either it or the Species-at-Risk policy will fail. The recovery plan was published in 2012, and a formal five-year review of caribou range recovery is due by 2017. A multibillion-dollar high-stakes showdown is looming between provinces and the federal government. Given the huge economic stakes, the risk of rewriting Species-at-Risk legislation cannot be overstated. In this confrontational and controversial setting, a retrospective policy work-around will be challenging to achieve without completely rewriting recovery goals to include economic feasibility. Regardless of these retrospective difficulties, the five protected populations in the north likely represent the best chance for caribou and boreal forest conservation in Alberta.

Sage-Grouse Protection in Western North America

The second case study of the greater sage-grouse illustrates the benefits of a proactive conservation planning approach to biodiversity conservation. In this case, the greater sage-grouse is an ideal flagship species for the interior sagebrush ecosystem of western North America.[16] Sage-grouse span a massive ecosystem across eleven western states (California, Colorado, Idaho, Montana, Nevada, North Dakota, Oregon, South Dakota, Utah, Washington, and Wyoming) and two Canadian provinces (Alberta and Saskatchewan). Grouse populations have declined range-wide from a variety of top-down threats from energy development and cultivation in the Rocky Mountain states, and catastrophic wildfire, invasive annual grasses, and conifer invasion across the Great Basin.[17] As a result of these threats and declines, greater sage-grouse were considered as a candidate species for listing under the Endangered Species Act (ESA). Strategic conservation planning kicked in early because user groups and state and federal governments wished to avoid the ESA listing and worked proactively and collaboratively to reduce threats.

Early regional planning showed that 75 percent of sage-grouse lived on 25 percent of the landscape in Wyoming.[18] This spatial prioritization provided the foundation for partners to target conservation efforts. They didn't have to save everything everywhere on every acre—triage provided flexibility to energy development while retaining enough habitat that conservationists could live with it.[19] While originally pioneered in Wyoming, this sort of spatial prioritization approach has expanded throughout all of sage-grouse range to include the other top-down threats and ultimately may be successful at recovering the sage-grouse across its range.[20]

In contrast to the caribou case study, the threat of ESA listing by the US Fish and Wildlife Service provided the motivation for a strategic landscape conservation approach rather than a retrospective punishment for violating legislation. Success in this case came by proactively reducing top-down threats that otherwise fragment sagebrush landscapes. As the primary land managers, the federal Bureau of Land Management and the US Forest Service completed the bulk of the public land policy protections. But, in contrast to the caribou case study, 30 percent of sage-grouse range is privately owned ranchlands. Thus, states and the federal Natural Resources Conservation Services–led Sage Grouse Initiative have invested nearly $0.5 billion to conserve and enhance an additional 5 million hectares of private lands inside sage-grouse strongholds. Through this proactive, coordinated, voluntary, and incentive-based effort, there is now a clear roadmap for landscape-scale conservation of at-risk species and the ecosystems they represent.

Moving from Best Practice to Common Practice

These two case studies illustrate the costs of waiting until it's too late to take a strategic landscape-scale perspective. In the case of caribou, economic costs of retrospective recovery to local, provincial, and federal governments may be so huge from a societal and economic viewpoint that it risks the entire Species-at-Risk Act legal framework. This is not a case of so-called new conservation giving in to industry and giving up on old conservation (protecting nature for its own sake).[21] This is simply pragmatism in the face of the unachievable costs of protecting threatened species while recognizing socioeconomic reality. Without more effectively conserving habitat, within the next few decades Canada will likely lose more than half its present woodland caribou populations,[22] with continued degradation of the ecosystem services, biodiversity, and ecological integrity of Canada's entire boreal forest. In contrast, the sage-grouse case

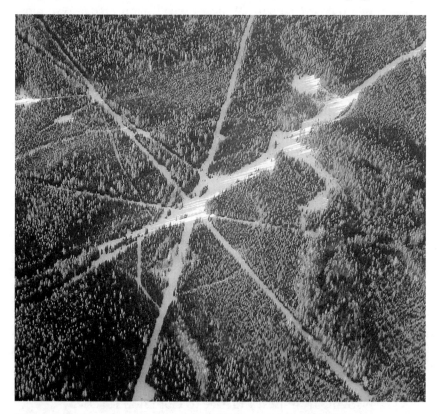

Figure 3-2. *Oil and gas seismic exploration lines and pipelines fragment woodland caribou habitat in the boreal forests of Alberta, facilitating access by carnivores such as wolves that leads to reduced habitat quality and population declines of this threatened species. Photo credit: Mark Hebblewhite.*

Figure 3-3. *Aerial view of the dense habitat disturbance created by numerous drilling pads and well heads for oil and gas development on the Jonah Field near Pinedale, Wyoming. Photo credit: Joseph M. Kiesecker.*

study illustrates how a proactive, coordinated, landscape-scale conservation approach can maximize biodiversity conservation outcomes and maintain energy development. The ultimate question now is how to get strategic planning for energy development with biodiversity conservation at the front of all development planning.

Figure 3-4. *Road mortality of sage-grouse. Increased oil and gas development results in increased traffic on roads. Photo credit: Jeremy R. Roberts, Conservation Media.*

Acknowledgments

For valuable discussions, I wish to thank colleagues Dave Hervieux, Stan Boutin, Marco Musiani, Nick Decesare, and I thank Brady Allred for data on wells drilled since 2012. Funding was provided by University of Montana and NASA ABoVE grant NNX15AW71A.

Notes

1. Andrew Nikiforuk, *Tar Sands: Dirty Oil and the Future of a Continent* (Vancouver, BC: Greystone Books, 2010).
2. James Hansen, "Game Over for the Climate," *New York Times*, May 9, 2012, 9.
3. Lisa Venier, Ian D. Thompson, et al., "Effects of Natural Resource Development on the Terrestrial Biodiversity of Canadian Boreal Forests," *Environmental Reviews* 22, no. 457 (2014), doi:10.1139/er-2013-0075.
4. David Schindler and P. G. Lee, "Comprehensive Conservation Planning to Protect Biodiversity and Ecosystem Services in Canadian Boreal Regions under a Warming Climate and Increasing Exploitation," *Biological Conservation* 143, no. 7 (2010): 1571–86, doi:10.1016/j.biocon.2010.04.003; Corey Bradshaw, Ian Warkentin, and Navjot Sodhi, "Urgent Preservation of Boreal Carbon Stocks and Biodiversity," *Trends in Ecology & Evolution* 24 (Oct. 2009): 541–48, doi:10.1016/j.tree.2009.03.019.
5. Orphe Bichet, Angelique Dupuch, Christian Hebert, Helene Le Borgne, Daniel Fortin, "Maintaining Animal Assemblages through Single-Species Management: The Case of Threatened Caribou in Boreal Forest," *Ecological Applications* 26, no. 2 (2016): 612–23.
6. Environment Canada, "Recovery Strategy for the Woodland Caribou (*Rangifer tarandus caribou*), Boreal population, in Canada," Species at Risk Act Recovery Strategy Series,Environment Canada Ottawa (2012), http://www.registrelep-sararegistry.gc.ca/default.asp?lang=En&n=33FF100B-1.
7. Dave Hervieux, Mark Hebblewhite, Nick DeCesare, Mike Russell, Kirby Smith, Sandi Robertson, and Stan Boutin, "Widespread Declines in Woodland Caribou (*Rangifer tarandus caribou*) Continue in Alberta," *Canadian Journal of Zoology* 91 (Oct. 2013): 872–82, doi:10.1139/cjz-2013-0123.
8. Environment Canada, "Scientific Assessment to Inform the Identification of Critical Habitat for Woodland Caribou (*Rangifer tarandus caribou*), Boreal Population, in Canada: 2011 Update," *Environment Canada Ottawa* (2011), http://www.registrelep-sararegistry.gc.ca/document/doc2248p/toc_tdm_st_caribou_e.cfm.
9. Dave Hervieux, Mark Hebblewhite, Dave Stepnisky, Michelle Bacon, and Stan Boutin, "Managing Wolves (*Canis lupus*) to Recover Threatened Woodland Caribou (*Rangifer tarandus caribou*) in Alberta," *Canadian Journal of Zoology* 92, no.12 (2014): 1029–37, doi:10.1139/cjz-2014-0142
10. Bob Weber, "Drilling Continues on Critical Alberta Caribou Habitat Despite Recovery Deadline," *Edmonton Journal*, April 16, 2015.
11. Environment Canada, "Scientific Assessment to Inform."
12. Richard Schneider, Grant Hauer, W.L. (Vic) Adarnowicz, and Stan Boutin, "Triage for Conserving Populations of Threatened Species: The Case of Woodland Caribou

in Alberta," *Biological Conservation,* 143, no. 7 (July 2010): 1603–11, doi:10.1016/j
.biocon.2010.04.002.

13. Eric Denhoff "Setting Alberta on the Path to Caribou Recovery," by Eric Denhoff,
    Mediator's Report, Alberta Environment and Parks, Government of Alberta (May 30,
    2016), http://esrd.alberta.ca/fish-wildlife/wildlife-management/caribou-manage-
    ment/caribou-action-range-planning/documents/OnThePathtoCaribouRecovery-
    May-2016.pdf.

14. Richard Schneider, Grant Hauer, Kimberley Dawe, W. L. (Vic) Adamowicz, and Stan
    Boutin, "Selection of Reserves for Woodland Caribou Using an Optimization Ap-
    proach," *PLOS One* 7, no. 2 (2012), doi:10.1371/journal.pone.0031672.

15. Denhoff, "Setting Alberta on the Path."

16. Steven Knick and John Connelly, eds., "Greater Sage-Grouse: Ecology and Conserva-
    tion of a Landscape Species and Its Habitats," *Studies in Avian Biology* (Berkeley: Uni-
    versity of California Press, 2011); Holly Copeland, "Conserving Migratory Mule Deer
    through the Umbrella of Sage-Grouse," *Ecosphere* 5, no. 9 (2014): 1–16, doi:10.1890
    /ES14-00186.1.

17. Knick and Connelly, "Greater Sage-Grouse"

18. Kevin Doherty, David Naugle, Holly Copeland, Amy Pocewicz, and Joseph Kiesecker,
    "Energy Development and Conservation Tradeoffs: Systematic Planning for Sage-
    Grouse in Their Eastern Range," in *Greater Sage-Grouse: Ecology and Conservation of a
    Landscape Species and Its Habitats,* edited by Steven Knick and John Connelly (Berke-
    ley: University of California Press, 2011); Holly Copeland, Amy Pocewicz, David
    Naugle, Tim Griffiths, Doug Keinath, Jeffrey Evans, and James Platt, "Measuring the
    Effectiveness of Conservation: A Novel Framework to Quantify the Benefits of Sage-
    Grouse Conservation Policy and Easements in Wyoming," *Plos One* 8 (June 2013),
    doi:10.1371/journal.pone.0067261.

19. David Naugle, ed., *Energy Development & Wildlife Conservation in Western North America.*
    (New York: Island Press, 2010).

20. Kevin Doherty, David Naugle, and Jeffrey Evans, "A Currency for Offsetting Energy
    Development Impacts: Horse-Trading Sage-Grouse on the Open Market," *Plos One* 5
    (April 2010), doi:10.1371/journal.pone.0010339.

21. Georgina Mace, "Whose Conservation," *Science* 345, no. 6204 (2014): 1558–

60. doi:10.1126/science.1254704; Emma Marris, "New Conservation Is an Expansion of
    Approaches, Not an Ethical Orientation, *Animal Conservation* 17, no. 6 (2014): 516–17,
    doi:10.1111/acv.12129.

22. Environment Canada, "Scientific Assessment to Inform."

CHAPTER 4

Win-Win for Wind and Wildlife

Joseph M. Kiesecker, Jeffrey S. Evans, Kei Sochi, Joe Fargione,
Dave Naugle, and Kevin Doherty

Throughout the world, countries are planning how to satisfy growing energy demands and, in the shadow of an increasingly changing climate, how to advance alternatives to fossil fuels. The International Energy Agency's 2040 forecast predicts renewable energy generation reaching 17,970 terawatt hours (51 percent of global electricity demand), with a significant portion of that coming from a tenfold increase in wind energy. In the United States, the world's largest cumulative producer of greenhouse gases, federal and state renewable energy policies moved forward rapidly, culminating in the Department of Energy's (DOE) vision for 35 percent of the United States's electricity generation from wind by 2050—we'll refer to this as the "DOE vision."[1]

At first glance, wind energy appears to be an environmentally friendly alternative to fossil fuels. In practice, however, wind presents some challenges. Wind energy has, per unit energy, a larger terrestrial footprint than most other forms of energy production[2] and has adverse impacts for wildlife.[3] New wind farms that could deliver the approximately 404 gigawatts of onshore wind energy needed to meet the DOE's vision have a predicted impact to approximately 7 million hectares of land, an area roughly the size of North Dakota. Additionally, these new farms would require an additional approximately 47,000 kilometers of new transmission lines.[4]

Furthermore, because wind development often occurs in concert with other forms of energy development (e.g., oil and gas) and other forms of land conversion (e.g., housing), the broad footprint of these farms means they'll likely lead to land-use conflicts. These conflicts could be reduced with appropriate siting that avoids areas of high conservation value. Over time, the accumulation of individual and disconnected decisions will have a great impact and pose the greatest challenge to the goal of protecting the environment while expanding energy development. If land planners comprehensively viewed the big picture—years in advance of energy development—they could identify and help avoid conflicts that pit development needs against the value and long-term functional health of other natural resources.

Achieving Wind Energy and Conservation Goals

In this chapter we look at two US-based case studies that facilitate wind energy expansion while reducing impacts to biodiversity and water resources. First, we highlight how and where nationwide DOE wind energy forecasts for individual states can be met within landscapes already disturbed.[5] Second, we underscore the value of scenario building to help reduce cumulative impacts to drinking water and aquatic ecosystems resulting from wind and shale gas development in the Marcellus Shale gas play.[6]

Repowering with Wind

The DOE vision outlines a spatial and temporal road map for meeting wind energy goals with specific projections for each state in the contiguous United States. To estimate the potential for wind energy generation on disturbed lands, we estimated the land area needed to meet wind energy scenarios within each state for the first 241 gigawatts (GW) needed by 2030.[7] Land area required to meet DOE vision depends largely on the minimum necessary wind potential (wind power class) of any given area.[8] Our built-out findings show that a network of land-based turbines, accounting for areas inappropriate for their placement, has the potential to generate 7,705 GW in total across the lower forty-eight states, with potential for

3,554 GW in areas already disturbed by human activities.[9] With a DOE projection of 404 terrestrial GW required by 2050, the United States can clearly reap the benefits of wind energy production in disturbed areas, which are also less likely to have high wildlife values.[10]

Only nine states (California, Arizona, Nevada, Utah, West Virginia, Pennsylvania, Virginia, North Carolina, and Tennessee) would be unable to meet DOE projections wholly within already disturbed areas. Furthermore, because of the uneven distribution of wind power classes, an additional nine states (Colorado, Idaho, Montana, New York, Oklahoma, Oregon, South Dakota, Washington, and Wyoming) would require an increased land base to generate the same amount of gigawatts if development is focused solely on disturbed lands. Notwithstanding these trade-offs, a disturbance-focused development strategy could avert the conversion of approximately 2.3 million hectares of undisturbed lands relative to an unconstrained scenario in which development is guided solely on the principle of maximizing wind potential (fig. 4-1).[11]

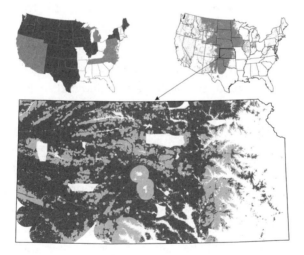

Figure 4-1. *Map of continental United States with states where DOE targets can (black) and cannot (gray) be met on disturbed lands, focusing on the 31 states that comprise the majority of the DOE vision, excluding states (white) with less than one Gigawatt of projected development; right-side map shows economically viable wind resources. Inset map of Kansas showing areas suitable for wind (black) and recommended avoidance areas (gray), adapted from Obermeyer et al., 2011.*

Assessing Environmental Impacts

The largest shale gas deposits in the United States are in the Marcellus Shale gas play, covering approximately 160,934 square kilometers across eight states and centered in West Virginia, Pennsylvania, and New York.[12] This area is also home to vast mountain ridgetops that provide some of the best terrestrial wind energy sites near the largest population centers in the United States. With the potential for wind plus oil and gas development come the challenges of land-use changes such as deforestation and expansions in impervious surfaces.[13] Aquatic ecosystems are particularly vulnerable to these developments. Deforestation and increases in impervious surfaces influence sediment, hydrologic, and nutrient regimes, which in turn influence aquatic biota and ecological processes.[14] This is of heightened interest in the Marcellus Shale gas play as it overlaps critical watersheds that provide drinking water for more than 22 million people in several of the largest metropolitan areas in the eastern United States (New York City, Philadelphia, and Washington, DC; fig. 4-2). To examine the link between energy development, deforestation, and water quality, we developed build-out scenarios for future energy development to quantify potential impacts to surface drinking water resources.[15]

First, we created predictive surfaces for wind and shale gas development for portions of states in the Central Appalachians. Second, we modeled build-out scenarios, guided by published projections from federal land management agencies.[16] And finally, we evaluated the effects of sequentially intensive build-out scenarios by measuring impacts to watershed health and identifying areas important for drinking water that are particularly vulnerable to future development.[17] Our findings suggest that to fully exploit the resource, up to 106,004 new gas wells and 10,798 new wind turbines would need to be built.[18] This would affect roughly 495,357 hectares of forest land and lead to more than 603,278 hectares of additional impervious surfaces, an area larger than the state of Delaware.[19] In order to reap the wind energy benefits, the states and regions must also figure out how to mitigate these impacts.

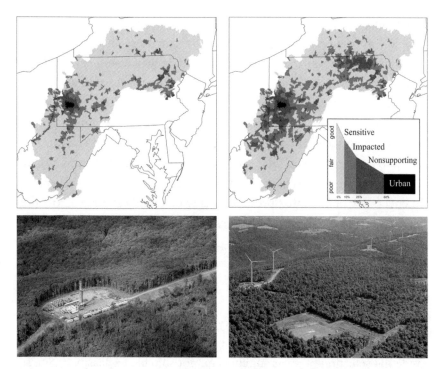

Figure 4-2. *Classification of percentage of impervious surface for 2006 predevelopment and full-build-out scenario, illustrating increased impervious cover and resulting risk to watershed health and surface water quality. Twenty-two million people are dependent on surface water to supply drinking water in this area. Photographs of shale gas and wind farm footprints.*

Resolving Trade-Offs. Shifting energy production from fossil fuels to renewable energy, which uses a broader spatial area and pulls from more diffuse sources, requires trade-offs that are not easily resolved. On the one hand, greater reliance on wind energy promises reduced CO_2 emissions and reduced water demand for electricity generation.[20] On the other hand, wind energy causes habitat loss and fragmentation of landscapes; that land loss will be difficult to overcome or mitigate against.[21] It makes sense to put new wind facilities on converted land areas that are low-quality habitats and already altered to such an extent that they can no longer viably support natural communities.[22] New wind development would have less impact on terrestrial wildlife if sited in these disturbed areas.

Additionally, given the nationwide surplus in wind energy in the United States, it makes sense to enable states that cannot meet wind energy production goals on disturbed lands to import electricity from states where there is a surplus of disturbance-based wind energy. A similar process could facilitate development globally—with countries that cannot meet renewable energy goals on disturbed lands importing electricity from countries where there is a surplus of disturbance-based renewable energy. To enact new policies that promote this wind energy sharing across borders, planners need to identify *go areas* at both regional and national levels. These identifications will highlight potential conflicts and opportunities for interstate or intercountry cooperation and address concerns that show up at the local level. Figure 4-1 shows what this type of planning might look like.

Potential impacts on biodiversity and surface water supply must be researched before development occurs, to help mitigate conflicts and to guide the selection of areas suitable and not suitable for leasing.[23] Site selection obviously requires coordination across jurisdictional boundaries, but given the potential for cumulative impacts, it is likely worth the effort (see chap. 8 for an example). Investments in policy, planning, and stakeholder processes can also help to reduce a project's risks for developers and financiers. In light of projected rapid growth in wind and shale gas, as well as other forms of energy development, our efforts to forecast development patterns and impacts in the Marcellus could serve as a model for thinking about development more broadly. Already Argentina, Australia, China, Colombia, and South Africa have identified large shale gas deposits that are in the early planning stages of development.

While impacts from individual gas wells and wind turbines or even those of a single wind farm or gas field are likely to be manageable and compatible with broader landscape-level conservation goals, the cumulative impacts of all these developments is a challenge.[24] Despite this, assessments of environmental impacts continue to be made well by well or gas field by gas field with little or no attempt to assess cumulative impacts of development. Scenario analyses such as ours from the Marcellus Shale gas play could help land planners and policy makers meet energy development goals more sustainably.[25] However, scenario-building approaches are rarely employed, at least in any formal way, in environmental impact assessments (EIAs). Widespread use of scenario modeling would enable regulators to examine potential consequences of development objectives more comprehensively, quickly, and inexpensively.

Making Wind Energy Work

The 2015 Paris Agreement is a call to arms for countries to reduce greenhouse gas emissions, and to do so in no small part by turning to renewable sources of energy. There are enormous challenges to implementing a large-scale wind energy strategy, including the technological obstacles of infrastructure and reliability issues. More critically, there are environmental issues that are not easily ignored: wind energy has a larger land footprint per gigawatt than most other forms of energy production, making appropriate siting and mitigation essential.

The great irony of the push to develop renewable energy to alleviate impacts of climate change is the specter of the daunting negative impacts of development that threaten to offset the environmental gains. Guiding development toward converted land may represent the best opportunity to mitigate impacts associated with climate change.[26] Animals and ecological systems need large, intact natural habitats that they can easily move through in order to adapt to a changing climate.[27] The potential benefits to biodiversity from climate change mitigation will be realized only if renewable energy development can proceed while avoiding and mitigating impacts to critical remaining habitat.[28] We have shown that in the United States it is possible to develop wind energy resources and maintain these intact ecosystems.

We have the tools to more effectively forecast where and how development could take place in a multiuse and multiobjective landscape. Regional scenarios are the perfect framework for regulators, industry, and policy makers to use in order to proactively examine how energy development will impact their area. States could require such preplanning, or industry standards might call for it, because in the end, a comprehensive energy development strategy will reduce risk and eliminate the need for expensive mitigation efforts.

Acknowledgments

We thank Donna Heimiller, National Renewable Energy Laboratory, and Paul Cryan, U.S. Geological Survey, for help with study design and analysis, and Jan Slatts, Jim Platt, and Lynn Scharf for help with data preparation. The work was supported by grants from the American the Wind Wildlife Institute, the Robertson Foundation, the U.S. Department of Energy, and the World Wildlife Fund. *The findings and conclusions in this article are those of the author(s) and do not necessarily represent the views of the U.S. Fish and Wildlife Service.*

Figure 4-3. Aerial view of the Elk River Wind Project near the small town of Beaumont, in the southern Flint Hills region of Kansas. This 150 MW wind farm came online in December 2005. Photo credit: Jim Richardson.

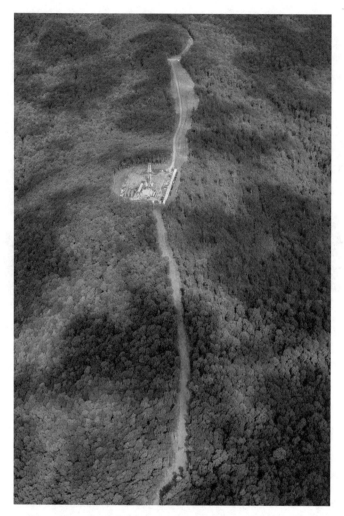

Figure 4-4. *Aerial photograph showing natural gas well site and fracking operation with related road and pipeline infrastructure in northeastern Pennsylvania forest. With new technology, energy companies are now able to extract natural gas from Marcellus Shale, a layer of black rock that runs beneath Pennsylvania, Ohio, Virginia, West Virginia, and New York. With the potential for upwards of 106,004 new wells and 10,798 new wind turbines resulting in up to 535,023 ha of impervious surface and upwards of 447,134 ha of impacted forest, mitigating the impacts of energy development will be one of the major challenges in the coming decades. Photo credit: Mark Godfrey, The Nature Conservancy.*

Figure 4-5. *A flock of starlings flies through a wind farm located in an agricultural field in California. While wind energy has a relatively large land footprint per unit of energy produced, impacts can be mitigated by siting wind farms in areas already disturbed, such as agricultural areas. Photo credit: Joseph M. Kiesecker.*

Notes

1.   U.S. Department of Energy, *Wind Vision: A New Era for Wind Energy in the United States,* Office of Energy Efficiency and Renewable Energy, March 2015, http://www.energy .gov/sites/prod/files/WindVision_Report_final.pdf.

2   McDonald et al., "Energy Sprawl" (see chap. 1, n. 4).

3.   Edward Arnett, W. Kent Brown, Wallace Erickson, Jenny Fiedler, Brenda Hamilton, Travis Henry, Aaftab Jain, Gregory Johnson, Jessica Kerns, Rolf Koford, Charles Nicholson, Timothy O'Connell, Martin Piorkowski, and Roger Tankersley Jr, "Patterns of Bat Fatalities at Wind Energy Facilities in North America," *Journal of Wildlife Management* 72, no. 1 (2008): 61–78, doi:10.2193/2007-221; Nathan Jones, Lisa Pejchar, and Joseph Kiesecker, "The Energy Footprint: How Oil, Natural Gas, and Wind Energy Affect Land for Biodiversity and the Flow of Ecosystem Services," *Bioscience* 65 no. 3 (March 2015): 290–301, doi:10.1093/biosci/biu224; Thomas Kunz, Sidney Gauthreaux, Nickolay Hristov, Jason Horn, Gareth Jones, Elisabeth Kalko, Robert Larkin, Gary McDracken, Sharon Swartz, Robert Srygley, Robert Dudley, John Westbrook, and Martin Wikelski, "Aeroecology: Probing and Modeling the Aerosphere," *Integrative and Comparative Biology* 48, no.1 (2008): 1–11, doi:10.1093/icb/icn037; National Research Council Committee, *Environmental Impacts of Wind Energy Projects* (Washington, DC: National Academies Press, 2007), doi:10.17226/11935.

4. US DOE, "Wind Vision."

5. Joseph Kiesecker, Jeffrey Evans, Joe Fargione, Kevin Doherty, Kerry Foresman, Thomas Kunz, Dave Naugle, Nathan Nibbelink, and Neal Niemuth, "Win-Win for Wind and Wildlife: A Vision to Facilitate Sustainable Development," *PLoS One* 6, no. 4 (2011), doi:10.1371/journal.pone.0017566.

6. Jeffrey Evans and Joseph Kiesecker, "Shale Gas, Wind and Water: Assessing the Potential Cumulative Impacts of Energy Development on Ecosystem Services within the Marcellus Play," *PLoS ONE* 9, no. 2 (2014), doi:10.1371/journal.pone.0089210.

7. Kiesecker et al., "Win-Win for Wind."

8. D. M. Heimiller and S. R. Haymes, *Geographic Information Systems in Support of Wind Energy Activities at NREL*(National Renewable Energy Laboratory, 2001).

9. Kiesecker et al., "Win-Win for Wind."

10. US DOE, "Wind Vision."; Kiesecker et al., "Win-Win for Wind."

11. Kiesecker et al., "Win-Win for Wind."

12. Evans and Kiesecker, "Shale Gas, Wind."

13. Jones et al., "The Energy Footprint."; Sally Entrekin, Michelle Evans-White, Brent Johnson, and Elisabeth Hagenbuch, "Rapid Expansion of Natural Gas Development Poses a Threat to Surface Waters," *Frontiers in Ecology and the Environment* 9, no. 9 (2011): 503–11, doi 10.1890/110053; Sujay Kaushal, Peter Groffman, Gene Likens, Kenneth Belt, William Stack, Victoria Kelly, Lawrence Band, and Gary Fisher, "Increased Salinization of Fresh Water in the Northeastern United States," *Proceedings of the National Academy of Sciences of the United States of America* 102, no. 38 (2005): 13517–20, doi:10.1073/pnas.0506414102; Daniel Rozell and Sheldon Reaven, "Water Pollution Risk Associated with Natural Gas Extraction from the Marcellus Shale," *Risk Analysis* 32, no. 8 (2012): 1382–93, doi:10.1111/j.1539-6924.2011.01757.x.

14. Thomas Cuffney, Robin Brightbill, Jason May, and Ian Waite, "Responses of Benthic Macroinvertebrates to Environmental Changes Associated with Urbanization in Nine Metropolitan Areas," *Ecological Applications* 20, no. 5 (2010): 1384–1401, doi:10.1890/08-1311.1.

15. U.S. Energy Information Administration, *Annual Energy Outlook 2013 with Projections to 2040* (2013), http://www.eia.gov/forecasts/aeo/pdf/0383(2013).pdf.

16. Ibid.

17. Collin Homer, Chengquan Huang, Limin Yang, Bruce Wylie, and Michael Coan, "Completion of the 2001 National Land Cover Database for the Conterminous United States," *Photogrammetric Engineering and Remote Sensing* 73, no. 4 (2007): 337–41; Sheila Olmstead, Lucija Muehlenbachs, Jhih-Shyang Shih, Ziyan Chu, and Alan Krupnick, "Shale Gas Development Impacts on Surface Water Quality in Pennsylvania," *Proceedings of the National Academy of Sciences of the United States of America* 110, no. 13 (2013): 4962–67, doi:10.1073/pnas.1213871110; Radisav Vidic, Susan Brantley, Julie Vandenbossche, David Yoxtheimer, and Jorge Abad, "Impact of Shale Gas Development on Regional Water Quality, *Science* 340, no. 6134 (2013), doi:10.1126/science.1235009.

18. Evans and Kiesecker, "Shale Gas, Wind."; US EIA, Annual Energy Outlook 2013.

19. Evans and Kiesecker, "Shale Gas, Wind."

20. US DOE, "Wind Vision."

21. Jones et al., "The Energy Footprint."

22. Robert Fletcher Jr., Bruce Robertson, Jason Evans, Patrick Doran, Janaki Alavalapati, and Douglas Schemske, "Biodiversity Conservation in the Era of Biofuels: Risks and Opportunities," *Frontiers in Ecology and the Environment* 9, no. 3 (2011): 161–68, doi:10.1890/090091; Bruce Stein, Lynn Kutner, and Jonathan Adams, eds., *Precious Heritage: The Status of Biodiversity in the United States* (New York: Oxford University Press, 2000).

23. Kiesecker et al., "Development by Design" (see ch. 1, n. 5).

24. Ibid.

25. Ibid.

26. Jonathan Mawdsley, Robin O'Malley, and Dennis Ojima, "A Review of Climate-Change Adaptation Strategies for Wildlife Management and Biodiversity Conservation," *Conservation Biology* 23, no. 5 (2009): 1080–89, doi:10.1111/j.1523-1739.2009.01264.x.

27. Lee Hannah and Lara Hansen, pp. 329–42 in *Climate Change and Biodiversity*, edited by Thomas Lovejoy and Lee Hannah (New Haven, CT: Yale University Press, 2005).

28. Kiesecker et al., "Win-Win for Wind."; Kiesecker et al., "Development by Design" (see ch. 1, n. 5).

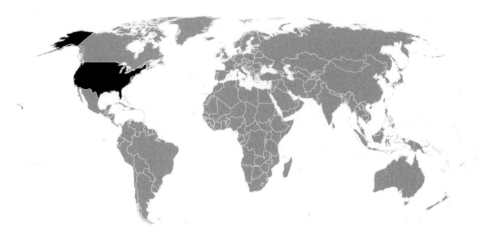

CHAPTER 5

Solar Energy Development and Regional Conservation Planning

D. Richard Cameron, Laura Crane,
Sophie S. Parker, and John M. Randall

Solar energy development has experienced rapid growth across the world, with a cumulative total of nearly 230 gigawatts (GW) of photovoltaic solar energy generating capacity installed by the end of 2015.[1] Utility-scale projects (those greater than 20 megawatts) have largely shifted toward photovoltaic (PV) systems, away from an earlier preference for concentrated solar power technology owing to the rapid decline in the cost of photovoltaic technology. Storing the energy generated by PV is a challenge because cost-effective battery technology is still in its early stages of development. Despite this limitation, PV has its advantages: it can generate electricity in a broader range of conditions than concentrated solar and is highly scalable.

The United States led the world in solar energy generating capacity until 1996, when Japan took the lead for a decade, only to be overtaken by Germany in the mid-2000s. China took over global leadership in PV capacity in late 2015. As of 2015, the most recent year for which figures are available, China's installed PV capacity was 43.5 GW; Germany's was 39.7 GW; Japan's was 34.4 GW; and that of the United States was 25.6 GW.[2]

Unrealized Potential

Despite this growth, the planning and permitting of large-scale solar facilities in the western United States has been slower than expected over the last ten years, owing in part to conflicts over what land is suitable for development.[3] Recently, some policies and incentives have directed siting to areas that have fewer conflicts with resource values, such as recreation and habitat conservation. In this chapter, we describe work in California deserts that has contributed to this progress and that may serve as a model for other types of infrastructure development.

A Call for Landscape Planning

In the mid-2000s, several factors converged to drive renewable energy development in the California deserts. Primary among these was the state of California's Renewables Portfolio Standard, which initially required utilities to get 20 percent of their energy from renewables by 2010, then increased the requirement to 33 percent by 2020, and was again raised in 2015 setting a goal of 50 percent of the state's energy by 2030.

At the same time, multiple federal policies drove a rapid increase in investment in the development of renewables, including the Energy Policy Act of 2005, which called for 10,000 megawatts of non-hydro renewable energy development on public lands by 2015. In addition, the 2009 American Recovery and Reinvestment Act included $43 billion for renewable energy development.

Even while policies and funding lined up, project permitting was challenging owing to several factors. First, public land management and regulatory agencies were overwhelmed by the number of project applications. Many of the projects proposed on public lands were submitted by speculators angling for investors by receiving an advantageous spot in the project review queue.[4] For many of the serious bids, project locations were chosen for access to transmission capacity and related financial factors without considering impacts on natural resources, such as endangered species, or other land uses. As such, many proposals ran into objections from environmental groups, the military (concerns about impacts to military training), or regulatory agencies, slowing down development time lines and jeopardizing developers' eligibility for stimulus funding and other tax credits.

Another limiting factor in the rate of project approvals was the lack of a landscape-level vision to balance energy development, resource protection, and other land uses. Without a comprehensive assessment and the tools to screen for potential impacts efficiently, agencies and stakeholders had to review potential impacts on a case-by-case basis. This made it difficult to rationalize which projects should be prioritized in permitting and development. Because the priorities for review were set based on application submission date, there was no incentive for developers to select areas with fewer environmental impacts.

California Answers the Call

The first effort to develop a comprehensive plan was the Bureau of Land Management's Solar Programmatic Environmental Impact Statement process that started in 2008. Also, in 2009, state and federal agencies initiated the Desert Renewable Energy Conservation Plan, an ambitious planning process to plan for both conservation and energy production across 22.5 million acres of public and private land in the California deserts. Recognizing that the existing public processes would take years and that there was a need to provide biodiversity conservation information in the interim, The Nature Conservancy began a planning effort to set a long-term vision of conservation success in the Mojave Desert ecoregion, a 32-million-acre area covering parts of four states and including much of the area covered by the Plan.

Traditionally, a conservation vision would identify a portfolio of lands that best represent the landscape's contribution to global biodiversity.[5] However, a vision for conservation success that maps priorities in a portion of the landscape doesn't inform land-use decisions in the "white space" outside these areas. In relatively intact landscapes such as the Mojave Desert, selecting a subset of areas to meet conservation targets leaves many intact and ecologically valuable areas in that white space. Looking at such a map, a developer or regulator might assume that these areas are suitable for development when, in fact, many of them are not. To inform project siting, it is critical for an assessment to characterize conservation value across all lands in the ecoregion, leaving no white space.

To address this, the Conservancy's 2010 Mojave Desert ecoregional assessment defined a vision of conservation success that can inform renewable energy siting.[6] This assessment used methods to characterize conservation value across the whole ecoregion, sometimes referred to as a "wall-to-wall" assessment. This approach was modeled after two other regional conservation assessments that had been completed for areas of Southern California and northern Baja California.[7] Results of the 2010 Mojave Assessment can be used to avoid piecemeal project impact evaluations that lead to a fragmented and degraded ecosystem. Additionally, it identifies already-degraded lands as a good first place to look for potential development areas.

Regional Planning Takes Shape

The Mojave Desert ecoregional assessment identifies lands that are important for long-term conservation of biodiversity. The highest-priority areas (areas of greatest conservation value) are divided into two categories: Ecologically Core and Ecologically Intact, based on how important an area is to meet quantitative conservation goals. The remaining lands are assessed for land cover disturbance and assigned to one of two lower-conservation-value classes: Moderately Degraded and Highly Converted (fig. 5-1). This comprehensive approach is meant to determine the relative suitability of land for conservation, not to find areas that are developable.

Using this assessment of conservation value, we conducted a solar energy trade-off analysis to see how much of the ecoregion is potentially suitable for solar development.[8] This study uses criteria important for utility-scale solar development, including land surface slope, solar insolation, and conservation status of land (e.g., excluding lands that preclude energy development, such as national parks). We found that as much as seven times the energy needed to meet the 33 percent California Renewables Portfolio Standard target could be attained on suitable Moderately Degraded or Highly Converted lands. Conversely, nearly 4 million acres

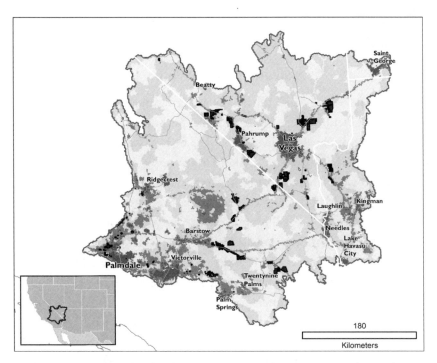

Figure 5-1. *Proposed solar development projects and conservation value in the Mojave Desert ecoregion. Conservation value is shown in four colors of gray, from light to dark in the following order: Ecologically Core, Ecologically Intact, Moderately Degraded, and Highly Converted. Solar development projects proposed from 2009 through 2013 on BLM land in Arizona, Nevada, and California, and projects proposed on private land in the Antelope Valley, California, are shown in black. Private land project data were not available for other parts of the ecoregion. Without information about the conservation value of all lands within the ecoregion, stakeholders could not evaluate the potential landscape-scale impact of proposed projects. Not all of these projects were built, and many have ceased.*

of Ecologically Core and Ecologically Intact lands met the slope, insolation, and land status requirements for development, highlighting that as much as 14 percent of the ecoregion's most ecologically valuable land is at risk for potential future development. This explicit characterization of the trade-offs between solar development and conservation in the Mojave Desert gives stakeholders and decision makers the data to move conversations beyond gridlock and has been held up as a rational, science-based approach to avoid conflict.

Solutions for the Future

As the global energy mix moves toward renewable energy sources, the case study described here can be a model for how to minimize trade-offs with conservation of biodiversity. Two key innovations from this work that can help shape decision making in contested landscapes include (1) the wall-to-wall assessment of conservation value and (2) the solar energy-specific trade-off analyses using the ecoregional assessment. When stakeholders can look at the trade-off or synergy between conservation and development, they can make better, more informed decisions.[9]

Too often, conservation plans define only priority areas for conservation, leaving vast parts of the landscape unassessed and therefore vulnerable to development. Planners might assume that areas labeled as having low conservation value today may have the potential to have greater conservation value in the future, especially when those landscapes can be effectively restored. However, in regions with a high percentage of intact lands and where disturbed and degraded lands are difficult to restore, the mapping of areas with low conservation suitability can both inform development decisions and enable conservation to proceed more rapidly and at lower cost.

This approach to landscape-level planning has been effective in advocating for renewable energy development that avoids damage to areas with high conservation value, and the approach has helped shape conservation policies in California and beyond. In 2012, the BLM completed the Programmatic Environmental Impact Statement to guide solar development on lands it manages in six southwestern states. The Mojave Desert conservation data helped inform the consensus recommendations for the solar portion of the Statement from environmental groups, solar energy developers, and utilities. As a result, some proposed development zone boundaries were adjusted to avoid high conservation value land. At the state level, the data were also used to help define the boundaries of Development Focus Areas and proposed conservation areas in the Desert

Figure 5-2. *Aerial photo of the sunpower facility in Rosamond, California. Note wind turbines in the distance. Photo credit: Dave Lauridsen for The Nature Conservancy.*

Renewable Energy Conservation Plan, a portion of which—the California desert BLM Land Use Plan Amendment—is scheduled to be completed in late 2016. In addition, all three investor-owned utilities in California— Pacific Gas and Electric, Southern California Edison, and San Diego Gas and Electric—which together supply roughly 75 percent of the electric power in California, consult the conservation value data when evaluating whether to sign agreements to buy power from certain projects.

Acknowledgments

Several people have contributed to the body of work described in this chapter and have been instrumental in developing, testing, and applying the methods for comprehensive ecoregional assessments. In particular, we would like to acknowledge Brian Cohen, Jim Moore, and Scott Morrison for their critical contributions to the Mojave Desert Ecoregional Assessment. Additionally, Erica Brand, Bill Christian, Mark Kramer, Charlotte Pienkos, and Jay Ziegler were instrumental in helping to make the science assessment relevant for federal and state policy implementation.

Notes

1. SolarPower Europe, "Global Market Outlook for Solar Power 2015–2019," 2014; International Energy Agency Photovoltaic Power Systems Programme, "Snapshot of Global PV Markets 2015," Report IEA PVPS T1-29: 2016 (2016).
2. IEA "Snapshot of Global PV."
3 ⁴ Louis Sahagun, "Environmental Concerns Delay Solar Projects in California Desert," Los Angeles Times, Oct. 19, 2009.
4. Julie Cart, "Land Speculators See Silver Lining in Solar Projects," Los Angeles Times, Feb. 18, 2012.
5. Craig Groves, Deborah Jensen, Laura Valutis, Kent Redford, Mark Shaffer, J. Michael Scott, Jeffrey Baumgartner, Jonathan Higgins, Michael Beck, and Mark Anderson, "Planning for Biodiversity Conservation: Putting Conservation Science into Practice," Bioscience 52, no. 6 (2002): 499–512, doi:0.1641/0006-3568(2002)052[0499:PFBCPC] 2.0.C.
6. John Randall, Sophie Parker, James Moore, Brian Cohen, Laura Crane, Bill Christian, Dick Cameron, Jason Mackenzie, Kirk Klausmeyer, and Scott Morrison, "Mojave Desert Ecoregional Assessment," unpublished report, The Nature Conservancy (2010).
7. M. D. White and Jerre Stallcup, "Las Californias Binational Conservation Initiative," Conservation Biology Institute (2004); Jerre Stallcup, "A Framework for Effective Conservation Management of the Sonoran Desert in California," Conservation Biology Institute (2009).

8. D. Richard Cameron, Brian Cohen, and Scott Morrison, "An Approach to Enhance the Conservation-Compatibility of Solar Energy Development," *PLoS One* 7, no. 6 (2012), doi:http://dx.doi.org/10.1371/journal.pone.0038437.

9. Kiesecker et al., "A Framework for Implementing" (see ch. 1, n. 26); Holly Copeland, Kevin Doherty, David Naugle, Amy Pocewicz, and Joseph Kiesecker, "Mapping Oil and Gas Development Potential in the US Intermountain West and Estimating Impacts to Species," *PLoS One* 4, no. 10 (2009), doi:10.1371/journal.pone.0007400.

CHAPTER 6

Planning for Offshore Oil

Eduardo Klein, Juan José Cardenas, Roger Martínez, Juan Carlos
González, Juan Papadakis, Kei Sochi, and Joseph M. Kiesecker

T he long shadow cast by British Petroleum's 2010 *Deepwater Horizon*
oil spill, perhaps the largest environmental disaster wrought by the
petroleum industry to date, brings into question the calculation of costs
against gains earned from offshore oil and gas drilling.[1] Truthfully, however, the question is moot. Today's industrial nations depend on oil. The
inevitable shift of balance away from fossil fuels toward renewables is
likely decades away. In the interim, fossil fuels will remain an important
part of the energy mix driving global development.[2]

Aided by advances in technology, oil producers are turning their attention to resources previously thought too difficult and expensive to tap:
oil and gas deposits in the oceans. Already more than a third of oil and
gas extracted worldwide comes from offshore sources. The International
Energy Agency estimates that forty-two percent of remaining recoverable
resources of conventional oil are located in offshore regions.[3] The most
productive areas are currently the North Sea, the Gulf of Mexico, the
Atlantic Ocean off the coast of Brazil, West Africa, the Arabian Gulf, and
seas off Southeast Asia.

Although public debate continues over the challenges of development
in deepwater marine environments, the continued exploitation of oceans
is unlikely to abate in the near term.[4] Public opinion might help stave
off impacts where explorations test the limits of technology, to mitigate

damage from potential accidents, and where explorations conflict with sensitive marine habitats or areas critical for sustaining livelihoods of local communities. In this chapter, we present a case study that examines how oil and gas development, specifically the thorny challenge posed by offshore development, can be guided to reduce impacts to marine ecosystems in the coastal environments along the Caribbean coast of Venezuela.

Offshore Oil in Venezuela

Venezuela is among the top ten biologically diverse countries in the world, and its southern Caribbean basin is a center of extraordinary marine biodiversity.[5] The basin harbors threatened species and is an important site for fishing, medicinal resources, and tourism. Owing to the demand and pressures placed on its resources, the basin is at high risk of overexploitation. Although Venezuela has marine protected areas in coastal waters, that protection covers only 3.4 percent of the country's waters, and, like most countries, Venezuela struggles to effectively manage these protected areas.

Alongside its rich biological diversity are Venezuela's abundant oil and gas reserves. These have made Venezuela the fifth largest oil exporter and seventh largest oil producer in the world.[6] In 2000, Venezuela developed a national plan for the exploration of petroleum resources in its continental platform, which is known to have more than 2.7 billion cubic meters of natural gas and 10 billion barrels of crude oil reserves. This plan is expected to affect more than 45 percent of the shallow maritime area (depths less than 200 meters) of the country. For a country like Venezuela, where more than 85 percent of the national income comes from oil, excluding exploration in offshore areas poses risks to the national economy and is neither entirely realistic nor desirable.

Petróleos de Venezuela, S.A. (PDVSA), a global energy corporation owned by the government of Venezuela, is responsible for managing and tapping those reserves. Its operations include the exploration, production, refining, transport, and marketing of hydrocarbons. The energy corporation has plans to divide the entire Venezuelan Caribbean Sea region into geographical blocks, which it will then lease to private energy companies that will be in charge of exploration and drilling. The private companies, however, will be obligated to follow the corporation's guidelines for development within each assigned block.

As part of this undertaking, the energy corporation signed an agreement with the Institute of Technology and Marine Science at Simón Bolivar University (INTECMAR) and The Nature Conservancy (TNC) to review information on the marine biodiversity in the Venezuelan Caribbean, to analyze the existing and potential threats to the marine ecosystems, and to establish a set of priority areas where conservation measures should be applied in order to preserve, for the long term, the marine environment. This initiative was continued in the Venezuelan Atlantic front with support from Chevron. In so doing, the marine science institute and The Nature Conservancy created a framework to guide the energy corporation's environmental licensing process and a set of recommended practices that could be used by the energy industry to reduce impacts on ecologically important areas identified for conservation.

For both the Caribbean and Atlantic Front projects, The Nature Conservancy applied a planning process used around the world to balance development objectives and conservation goals.[7] TNC has used this approach to classify a protected-areas network off-limits to mining in Mongolia[8] and to identify no-take zones in the Great Barrier Reef of Australia.[9] The process considers three primary questions and outputs to guide subsequent actions:

- What resources and functions within the landscape or seascape are critical to people and how much is needed to conserve the long-term viability of these natural resources? Resulting output includes a spatial blueprint of priority sites for conservation.

- How might cumulative development activity—from energy, fisheries, aquaculture, recreation, and other sectors—affect these resources over time? Outputs include maps indicating potential conflicts between development and conservation priorities and the opportunity to redesign either conservation areas or development objectives to reduce conflict.

- What opportunities and strategies are possible for addressing trade-offs and improving economic, social, and environmental outcomes? Outputs are a set of conservation-based standards and practices to be incorporated into best-management practices and mitigation requirements allowing development to advance in ways considerate of conservation goals.

While this approach has been applied in numerous places around the world, most applications have been in the terrestrial environment; while robust, this type of planning often does not address industrial development in the marine environment.[10]

Planning for Conservation and Development

We conducted a comprehensive regional assessment to evaluate the full spectrum of biodiversity in Venezuela's Caribbean waters. The assessment identified areas of biological significance that should be targeted for conservation and developed a framework for assessing impacts to marine resources from human activities. The area of study is 165,000 square kilometers of coastal and marine ecosystems in the Venezuelan Caribbean on the continental shelf (depths less than 200 meters). The region is divided into thirteen ecoregions and is home to thirty conservation targets that include deepwater coral communities, sea turtle communities, and mangrove forests. Ecological and biological experts determined key attributes for each target, including their existing condition (size, landscape context, and viability), vulnerability, and rarity, which were then used in determining conservation status. Experts selected indicators to be used by energy companies for quantifying the status of each conservation target. It was critical to establish conservation goals for each conservation target so we could estimate the area required to conserve in order to maintain long-term viability. This critical step can help determine when specific actions (e.g., developing an oil drilling platform) are consistent with conservation goals and when they are not. However, it is difficult to estimate these values in marine communities, where information is scarce and distributions are very widespread (e.g., habitat for whales, fish, or turtles). To obtain estimates, the group convened a series of expert workshops.

With additional expert consultation, we identified seven main human uses in the region and mapped their potential impacts to the conservation targets (e.g., oil exploration, river discharges, commercial fishing, coastal urban development, aquaculture farms, and maritime transport). The marine institute scientists selected priority conservation areas and, in collaboration with the energy corporation, we produced a final set of twenty priority sites representing more than 44,000 square kilometers and more than 38 percent of all marine areas on the continental shelf in the Caribbean, as well as nine areas in the Atlantic front covering 17,500 square kilometers and 46 percent of the marine space (fig. 6-1).

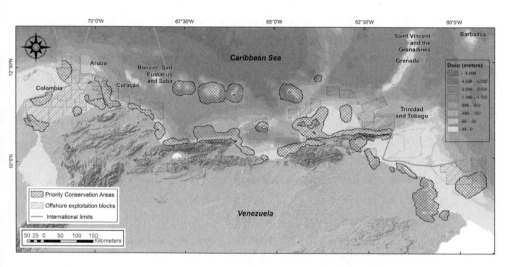

Figure 6-1. Priority conservation areas and offshore oil exploitation blocks in Venezuela's Caribbean and Atlantic front. The conservation areas selected contain a representation of the coastal and marine biodiversity conservation targets (species, communities, and ecosystems) that have the best conservation status with a minimum risk from natural factors (such as climate change) and anthropogenic activities (such as fisheries, maritime transport, and oil industry–related activities). (Prepared by J. Papadakis, Remote Sensing and Geospatial Analysis Laboratory, Universidad Simón Bolívar.)

A number of the priority conservation areas conflict with anticipated oil and gas exploitation blocks. For these sites, we attempted to predict the specific infrastructure that would be used, such as the construction of platforms, wells, and pipelines. We ran scenarios for various sites to highlight potential impacts and to articulate guidelines and environmental practices to the energy corporation. In addition, we created an oil impact best-practice matrix to provide detailed guidelines and practices for each stage of exploration and exploitation. The suggested guidelines aim to minimize impacts to marine biodiversity and were provided to the energy corporation and offshore operators to guide exploration policy and activities.

The best evidence for the utility of this process is that results have been used in several decisions made by the oil industry. In 2009, a three-dimensional seismic assessment in the Gulf of Paria was planned according to proposed recommendations, limiting the power of the sound cannons, the time of the year, and the time of the day, consistent with known information about marine mammals, iconic species, and migratory birds found in that area.[11] Another innovative outcome was the design of a pipeline route in the Gulf of Venezuela. Initially, three routes were proposed by the oil company, including one that followed a path mostly on land, which was a less expensive alternative to a marine route (fig. 6-2). Using the current information on the locations and condition of the marine ecosystems, important biodiversity values, and priority conservation areas, the team was able to produce a map of the potential accumulated environmental impacts. The route with the least impact happened to be the longest and more expensive one, but the route avoided the main biodiversity areas, fishing grounds, and the main maritime transit lanes. To make a final decision, the company evaluated the cost of the proposed route against possible future environmental liabilities and selected the path using the principle of least-accumulated environmental impacts.[12]

Our efforts have also led to plans to create a zoning scheme for offshore oil and gas exploitation activities in Venezuela and serves as a compelling case for more broadly integrating development objectives and conservation goals in marine environments elsewhere in the world. The

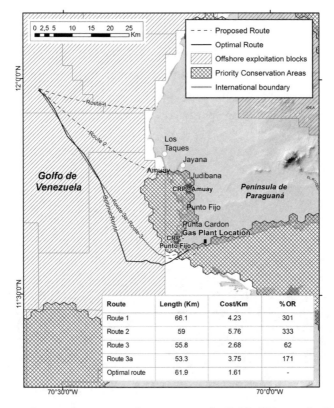

Figure 6-2. *Proposed marine pipeline routes in the Gulf of Venezuela. Route alternatives proposed by the industry and derived optimal route. The optimal solution was derived selecting the minimum accumulated environmental impacts along the route. (Prepared by J. Papadakis, Remote Sensing and Geospatial Analysis Laboratory, Universidad Simón Bolívar.)*

implementation of the proposed strategies and the monitoring of biological and ecological features could prevent irreversible damages to marine biodiversity and help mitigate unavoidable impacts.

Solutions Moving Forward

Through this partnership, The Nature Conservancy and the Institute of Technology and Marine Science at Simón Bolivar University have proposed a set of conservation-based standards and practices to be incorporated into the permits for each oil and gas lease. These cover potential development scenarios for oil development in each block, information on possible locations where infrastructure would be compatible with conservation goals and areas where conservation investments should be a priority. Outputs include a detailed management plan required to be implemented by future operators. In addition to government enforcement of environmental laws, the energy corporation, Petróleos de Venezuela,

Figure 6-3. A natural gas drilling platform at the mouth of Mobile Bay, Alabama, with an extension of Dauphin Island showing brown pelicans and terns in the foreground. Photo credit: Carlton Ward Jr.

will include these practices in each future contract associated with leasing of any block for development. This partnership between the energy corporation, the marine science institute, and The Nature Conservancy is an example of how collaborations between conservation organizations, petroleum developers, and environmental regulators can work.

Timing is key in similar collaborations. Here, we identified priority conservation areas and recommended practices at the beginning of the project life cycle. When project managers first understand biodiversity and important ecosystem values and develop recommended practices to minimize impacts of proposed development before that development occurs, two important outcomes result. First, energy corporations like Petróleos de Venezuela can avoid impacts on the most valuable ecological resources, and where development must proceed they can include guidelines, policies, and requirements for environmentally conscious practices

Figure 6-4. *Aerial photo of Port Fouchon, Louisiana, a major hub for servicing oil and gas rigs in the northern Gulf of Mexico. Although drilling platforms reside in the marine environment, typically offshore oil and gas development is supported with onshore infrastructure that also needs to be carefully sited to reduce impacts. Photo credit: Carlton Ward Jr.*

in contractual documents to minimize the likelihood of negative ecological impacts. Second, the energy company can help all parties involved to more effectively manage the environmental and economic risks in development. This guides decision making in such a way as to optimize the achievement of the multiple objectives in working landscapes and seascapes before ecological values are irretrievably lost, before livelihoods are profoundly and negatively impacted, and while there is still room to maneuver.

Notes

1. William Freudenburg and Robert Gramling, *Blowout in the Gulf: The BP Oil Spill Disaster and the Future of Energy in America* (Cambridge, MA, and London: MIT Press, 2012).
2. US EIA, Annual Energy Outlook 2013 (see ch. 4, n. 15).
3. Ibid.
4. Freudenburg and Gramling, *Blowout in the Gulf*; US EIA, Annual Energy Outlook 2013 (see ch. 4, n. 15); Donald Boesch and Nancy Rabalais, eds., *Long-Term Environmental Effects of Offshore Oil and Gas Development* (London and New York: Elsevier Applied Science, 1987); National Research Council, ed., *Oil in the Sea III: Inputs, Fates, and Effects* (Washington, DC: National Academy Press, 2003); Frode Olsgard and John Gray, "A Comprehensive Analysis of the Effects of Offshore Oil and Gas Exploration and Production on the Benthic Communities of the Norwegian Continental Shelf," *Marine Ecology Progress Series* 122 (1995): 277–306.
5. Patricia Miloslavich, Eduardo Klein, Edgard Yerena, and Alberto Martin, "Biodiversidad marina en Venezuela: Estado actual y perspectivas" ("Marine Biodiversity in Venezuela: Status and Perspectives"), *Gayana* 67, no. 2 (2003): 275–301; Eduardo Klein, *Prioridades de PDVSA en la conservación de la biodiversidad en el caribe venezolano* (Caracas, Venezuela: Petróleos de Venezuela, SA/Universidad Simón Bolívar/The Nature Conservancy, 2008), http://cbm.usb.ve/sv/prioridades-de-pdvsa-en-la-conservacion-de-la-biodiversidad-en-el-caribe-venezolano/; Eduardo Klein and Juan José Cárdenas, eds. *Identificación de la prioridades de conservación asociadas a los ecosistemas de las Fachada Atlántica y a su biodiversidad"* (Universidad Simón Bolívar–The Nature Conservancy, 2011), http://cbm.usb.ve/sv/identificacion-de-la-prioridades-de-conservacion-asocia das-a-los-ecosistemas-de-las-fachada-atlantica-y-a-su-biodiversidad/.
6. Klein, *Prioridades de PDVSA*; Klein and Cárdenas, *Identificación de la prioridades*.
7. Craig Groves, *Drafting a Conservation Blueprint: A Practitioner's Guide to Planning for Biodiversity* (Washington, DC: Island Press, 2003); Kiesecker et al., "Development by Design" (see ch. 1, n. 5).
8. Michael Heiner, Yunden Bayarjargal, Joseph Kiesecker, Davaa Galbadrakh, Nyamsuren Batsaikhan, Ganbaatar Munkhzul, Ichinkhorloo Odonchimeg, Oidov Enkhtuya, Donchinbuu Enkhbat, Henrik von Wehrden, Richard Reading, Kirk Olson, Rodney Jackson, Jeffrey Evans, Bruce McKenney, James Oakleaf, and Kei Sochi, "Identifying Conservation Priorities in the Face of Future Development: Applying Development by

Design in the Mongolian Gobi" (The Nature Conservancy, 2013), http://www.nature
.org/media/smart-development/development-by-design-gobi-english.pdf.

9. Leanne Fernandes, Jon Day, Adam Lewis, Suzanne Slegers, Brigid Kerrigan, Dan
    Breen, Darren Cameron, Belinda Jago, James Hall, Dave Lowe, James Innes, John
    Tanzer, Virgina Chadwick, Leanne Thompson, Kerrie Gorman, Mark Simmons,
    Bryony Barnett, Kirsti Sampson, Glenn De'ath, Bruce Mapstone, Helene Marsh,
    Hugh Possingham, Ian Ball, Trevor Ward, Kirstin Dobbs, James Aumend, Deb Slater,
    and Kate Stapleton, "Establishing Representative No-Take Areas in the Great Barrier
    Reef: Large-Scale Implementation of Theory on Marine Protected Areas," *Conservation
    Biology* 19, no. 6 (2005): 1733–44, doi:10.1111/j.1523-1739.2005.00302.x.

10. Melissa Foley, Benjamin Halpern, Fiorenza Micheli, Matthew Armsby, Margaret
    Caldwell, Caitlin Crain, Erin Prahler, Nicole Rohr, Deborah Sivas, Michael Beck, Mark
    Carr, Larry Crowder, J. Emmett Duffy, Sally Hacker, Karen McLeod, Stephen Palumbi,
    Charles Peterson, Helen Regan, Mary Ruckelshaus, Paul Sandjfer, and Robert Steneck,
    "Guiding Ecological Principles for Marine Spatial Planning," *Marine Policy* 34, no. 5
    (2010): 955–66.

11. Klein, *Prioridades de PDVSA*.

12. Klein and Cárdenas, *Identificación de la prioridades*.

CHAPTER 7

Energy and Ecosystem Services in Latin America

Heather Tallis

Three major trends are on a collision course in Latin America: growth in the energy sector, increasing national support for nature, and strong movements for equality and indigenous rights. Surprisingly, accounting for ecosystem services in energy development could help settle this building storm.

The Perfect Storm

Energy use is projected to grow 30 percent globally by 2040 and, in Latin America, that translates into a doubling or tripling of the current 1,300 terawatt hours generated on the continent by 2050.[1] Developing energy to meet these demands requires substantial new infrastructure. Peru alone needs to invest an estimated $33 billion in energy infrastructure over the next decade. While energy development is likely to buoy the economies of Latin American countries, it could have severe negative impacts on the environment.[2]

Major negative environmental outcomes would conflict with the bold steps several Latin American countries are taking to conserve nature. In 2008, Ecuador became the first country to recognize the rights of nature in its constitution, through a publicly supported referendum. Those rights include the right to exist, persist, and maintain and regenerate nature's vital cycles. Bolivia followed suit in 2010, passing a law that declares Mother Earth and life systems as inherent rights holders as laid out in the law.

An open question now stands as to how energy development in Latin America will balance energy needs with environmental impact and indigenous rights. This third element enters in when energy resources or their environmental impacts are in indigenous territories. Historically, energy development has not always respected indigenous rights, and indigenous groups have, in some cases, borne the bulk of environmental loss and associated human impacts.

The difficult challenge of balancing these values could be helped by passage of regulations or standards that address ecosystem services. Ecosystem services are benefits provided by nature such as clean drinking water, a stable climate, fertile soils, and places for spiritual moments. Energy development, if not done carefully, impacts these services and leads to damages like contaminated water, lost access to food or recreation, polluted air, and more. In the worst-case scenario, negative impacts accrue to people who are already marginalized, intensifying inequality. In that sense, energy development could cancel out its benefits or even become a net harm to society. If, however, ecosystem services are conserved through regulations or standards, people and nature can be protected, impacts minimized, and remaining impacts mitigated in a way that maintains equality.

Energy for the People

Unfortunately, in many Latin American countries, environmental regulations are just developing or do not yet exist.[3] Developing these regulations will require strong leadership and bold action. Here, we provide two examples of how incorporating ecosystem services into energy planning can lead to more balanced outcomes: coal mine permitting in Colombia and road planning in the Peruvian Amazon.

Colombia holds large coal reserves, especially in the northern Cesar Department, the leading coal-producing region in the country. In 2010, the Colombian Ministry of Environment and Sustainable Development adopted a resolution (MADS 2010—Resolución 1503 de 2010) and a methodology to improve the environmental licensing process for terrestrial projects.[4] However, by 2011, those laws did not yet call for consideration of ecosystem services. The Nature Conservancy and the Natural Capital Project engaged with the Colombian government in a conversation around emerging environmental regulations, including whether and how ecosystem services could be accounted for in mine project environmental impact assessments and mitigation considerations. As coal mining is projected to continue growing in the region, and many coal reserve areas intersect with indigenous territories, it is important to ensure that energy development does not have unintended adverse effects on people, especially indigenous populations.

During the same period, multilateral lenders expanded their thinking and began to include ecosystem services in their safeguard considerations. In 2012, the International Finance Corporation, a major lender for energy sector and infrastructure development, released new standards and guidelines that asked all lending recipients to consider ecosystem services in their proposed development project impacts.[5] The Inter-American Development Bank made an even bolder move and asked how its transport infrastructure loans could be at risk from ecosystem service impacts or loss during development.[6] This was a fundamentally different entry point than typical environmental safeguard considerations because the bank recognized the potential material risks to core business caused by unintentional harm to ecosystem services during infrastructure development.

Responding to these new standards and considerations, The Nature Conservancy and the Natural Capital Project conducted a second study on a proposed road project in the Peruvian Amazon to test how ecosystem service impacts could be accounted for in road development and planning. As roads are often developed in conjunction with expanding energy infrastructure, we see this application as broadly relevant to the energy growth context as well as to general infrastructure expansion.[7]

Servicesheds: An Ecosystem Services Impact Framework

Building off the long-established biodiversity mitigation hierarchy, we developed a framework to consider biodiversity and ecosystem services together in energy development and mitigation.[8] Here we focus on the "serviceshed" approach, as it is the most novel element of the framework and it directly addresses how people will be impacted by environmental changes associated with energy development.

A serviceshed is an area that provides a specific ecosystem service benefit to a specific group of people.[9] By drawing servicesheds in a given area, we can identify where and how energy development might reduce services, and we can see who will be impacted. Getting this information early in a development process helps private sector energy developers and government regulators know where development is likely to place people at greatest risk from service loss. Servicesheds also help identify where there are opportunities to mitigate lost services by improving ecosystems to provide the same benefits to the same people. Inequality is a critical consideration in these analyses. For example, one of the world's oldest wetland mitigation programs, guided by the US Clean Water Act, unintentionally created inequality. Wetlands were more often developed near urban areas, reducing the benefits those urban, often lower-income residents might get from clean wetlands, such as drinking water filtration and recreational fishing opportunities. These wetland losses were mitigated by creating or restoring wetlands in rural areas where people generally had higher incomes. In essence, by not using servicesheds, developers took wetland ecosystem services away from the urban poor and gave them to wealthier rural populations. This is clearly not the intent of the Clean Water Act, but this pattern emphasizes how easily environmental mitigation can create inequality if the connections between people and the environment are not closely tracked.

The serviceshed concept requires consideration of three components: supply of ecosystem services, physical access, and legal access to the service. For example, ecosystems might support abundant fish populations in lakes. But people must be able to access the supply, both physically and legally, for a benefit to flow. If the lakes with abundant fish are not connected to people by roads or walking paths, people do not have physical access to the fish. If the lakes are legally protected areas, people cannot legally harvest the fish. Servicesheds are drawn with these three elements in mind.

Comparing servicesheds with proposed energy development plans can reveal how many and which people are likely to lose ecosystem services.[10] In the Colombia mine development case, stakeholders were most concerned about mine impacts on downstream drinking water quality. Direct pollution from mining activities is already accounted for in other laws, so in this case, we focused on water quality impacts that could come from soil erosion and nutrient losses once trees and other vegetation were removed from the landscape. Given that most Colombians in the Cesar Department get their drinking water from municipal water supplies, the servicesheds for these water-related benefits are simply the watersheds upstream from municipal drinking water access points. We delineated these watersheds and overlaid them with proposed mine development areas (see the gray areas in fig. 7-1). Cities likely to lose drinking water quality are identified on our map with dots scaled to the population size of that city.

We used a similar approach in Peru, where drinking water quality impacts from proposed road development were also of concern.[11] Servicesheds for cities with municipal water delivery are delineated in the same way as in Colombia. However, indigenous populations mostly collect drinking water from surface streams and rivers near their villages. For their servicesheds, we mapped the nearest likely access points from villages and identified their upstream watersheds. As with the mines, overlaying the proposed road and the servicesheds identifies which indigenous and nonindigenous settlements are likely to be impacted (fig. 7-1c).

We also asked whether mitigation could equitably return the same benefits to the same people by modeling the potential for mitigation benefits within impacted servicesheds. This exercise reveals whether unintentional redistribution of ecosystem service benefits, such as that seen in the Clean Water Act in the United States, could be avoided.

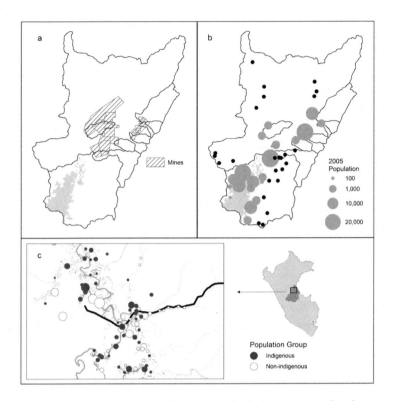

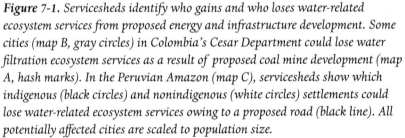

Figure 7-1. Servicesheds identify who gains and who loses water-related ecosystem services from proposed energy and infrastructure development. Some cities (map B, gray circles) in Colombia's Cesar Department could lose water filtration ecosystem services as a result of proposed coal mine development (map A, hash marks). In the Peruvian Amazon (map C), servicesheds show which indigenous (black circles) and nonindigenous (white circles) settlements could lose water-related ecosystem services owing to a proposed road (black line). All potentially affected cities are scaled to population size.

In the Colombia case, drinking water quality impacts could not be mitigated for four cities (fig. 7-2a). If coal mining proceeds as proposed, these four cities are vulnerable to ecosystem service loss, even if mitigation is required. The servicesheds are too small or too heavily developed for ecosystem service mitigation efforts to succeed. Mine expansion could be redesigned to avoid this potential service loss and creation of inequality among cities.

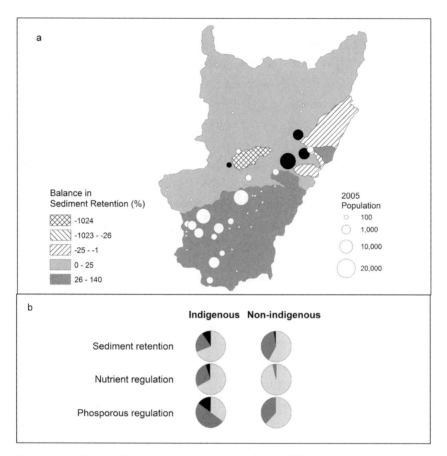

Figure 7-2. *Potential ecosystem service losers identified by energy and infrastructure impact and mitigation analyses within servicesheds. In both the Colombia (map A) and Peru (figure B) cases, some people are likely to suffer ecosystem service losses that cannot be mitigated. In Colombia, four cities (black circles) will have varying degrees of residual losses. In Peru, nearly all impacts to nonindigenous populations can be mitigated (gray pie sections), but indigenous settlements will face some drinking water filtration losses even after mitigation (black pie sections).*

We see a similar outcome in the Peru case, where the planned road is likely to affect indigenous people disproportionately, even after mitigation attempts (fig. 7-2b). Our results suggest that nearly all impacts to drinking water filtration can be mitigated for nonindigenous populations, but indigenous populations face higher sediment, nutrient, and phosphorous water pollution even after service mitigation. These results indicate that construction of the road as proposed may create ecosystem service losers even if ecosystem service mitigation is required, and indigenous people will likely suffer the biggest losses, which worsens existing inequalities. Altering the route of the road could avoid these negative impacts on both nature and people, and servicesheds could be a useful framework for the exploration of such alternatives.

From Best Practice to Common Practice

Precedent exists for the use of servicesheds as an approach for aligning energy development, ecosystem service provision, and equality. An increasing number of countries and lenders have development regulations that cover ecosystem services, but these are not enforced because methods and capacity to apply them are limited. The serviceshed method provides a replicable, robust way to inform energy development, identify how proposed development may impact people, and determine whether possible ecosystem service losses will create or worsen inequalities. As such, the concept closes the methods gap.

However, the capacity gap remains a challenge. For example, after we presented our findings in the Colombia case to the minister of environment, mines, and territorial development, he commented how useful the maps are in showing which cities are at risk and that this is exactly the kind of information his staff needed. However, he then emphasized that his staff did not have the capacity to run a model, making the kind of analyses we presented unlikely to occur. To apply the serviceshed concept in the thousands of energy development decisions that will be made in the next ten years requires rapid advances to make the method much more accessible. The Natural Capital Project has released a free, open-source software tool called OPAL that aims to fill this gap.[12] OPAL is specifically

Figure 7-3. View of coal extraction in the Cesar Valley of Colombia. While coal energy development has a small relative land footprint in terms of amount per unit energy, it has among the highest CO_2 emission of all energy types, and development creates severe impacts to wildlife habitat and creates issues for the maintenance of local water quality. Photo credit: Joseph M. Kiesecker.

Figure 7-4. Aerial view of mountaintop coal mining in West Virginia. Photo credit: Kent Mason.

designed for estimation of ecosystem services in a development-and-mitigation context.

Energy development is a delicate balancing act—it improves lives, but without careful planning it can also unintentionally harm people through lost ecosystem services. Servicesheds proactively reveal these harms and allow land planners, developers, and policy makers to consider alternatives that let energy development advance rapidly without the creation of damaging inequalities.

Notes

1. Walter Vergara, Claudio Alatorre, and Leandro Alves, "Rethinking Our Energy Future," Inter-American Development Bank (2013).
2. Oakleaf et al., "A World at Risk" (see ch. 1, n. 6).
3. Ana Villarroya, Ana Christina Barros, and Joseph Kiesecker, "Policy Development for Environmental Licensing and Biodiversity Offsets in Latin America," *PLoS One* 9 (2014), doi:10.1371/journal.pone.0107144.
4. Saenz et al., "Development by Design in Colombia" (see ch. 1, n. 24).
5. International Finance Corporation, *Performance Standard 6: Biodiversity Conservation and Sustainable Management of Living Natural Resources* (2012).
6. Lisa Mandle, Heather Tallis, Leo Sotomayor, and Adrian Vogl, "Who Loses? Tracking Ecosystem Service Redistribution from Road Development and Mitigation in the Peruvian Amazon, *Frontiers in Ecology and the Environment* 13 (2015): 309–15.
7. Lisa Mandle, Benjamin Bryant, Mary Ruckelshaus, Davide Geneletti, Joseph Kiesecker, and Alexander Pfaff, "Entry Points for Considering Ecosystem Services within Infrastructure Planning: How to Integrate Conservation with Development in Order to Aid Them Both," *Conservation Letters* 9, no. 3 (2016): 221–27, doi:10.1111/conl.12201/.
8. Heather Tallis, Christina Kennedy, Mary Ruckelshaus, Josh Goldstein, and Joseph Kiesecker, "Mitigation for One and All: An Integrated Framework for Mitigation of Development Impacts on Biodiversity and Ecosystem Services, *Environmental Impact Assessment Review* 55 (2015): 21–34, doi:10.1016/j.eiar.2015.06.005.
9. Heather Tallis and Steve Polasky, "Mapping and Valuing Ecosystem Services as an Approach for Conservation and Natural-Resource Management," *Annals of the New York Academy of Sciences* 1162 (2009): 265–83, doi:0.1111/j.1749-6632.2009.04152.x.
10. Heather Tallis and Stacie Wolny, "Including Ecosystem Services in Mitigation," Report to the Colombian Ministry of the Environment, Mines and Territorial Development (Stanford, CA: Natural Capital Project, 2010).
11. Mandle et al., "Who Loses? Tracking Ecosystem Service Distribution."
12. Lisa Mandle, James Douglass, Juan Sebastian Lozano, Richard Sharp, Adrian Vogl, Douglas Denu, Tomas Walschburger, and Heather Tallis, "OPAL: An Open-Source Software Tool for Integrating Biodiversity and Ecosystem Services into Impact Assessment and Mitigation Decisions," *Environmental Modelling and Software* 84 (2013): 121–33, doi:10.1016/j.envsoft.2016.06.008.

CHAPTER 8

Biofuels Expansion and Environmental Quality in Brazil

Christina M. Kennedy, Peter L. Hawthorne,
Kei Sochi, Daniela A. Miteva, Leandro Baumgarten,
Elizabeth M. Uhlhorn, and Joseph M. Kiesecker

Rising energy demands, volatile oil prices, and concerns about climate change have led countries to look for alternatives to fossil fuels.[1] Biofuels, or fuels produced from organic matter, have been embraced as a promising alternative to oil, because in principle they can lower carbon emissions, enhance domestic energy security, and revitalize rural economies.[2] More than sixty countries have biofuel targets or mandates, which have led global production to grow from 16 to 120 billion liters over the last decade.[3]

Most commercial production of biofuels comes from sources like starch and sugar-based ethanol, mainly produced in the United States and Brazil from corn or sugarcane; or, to a lesser extent, from plant oil–based biodiesel, mainly produced in Europe from rapeseed but also in Brazil from soybean and in Indonesia and Malaysia from oil palm.[4] These biofuel feedstocks can use three orders of magnitude more land per unit energy than natural gas, oil, coal, wind, hydropower, or solar.

Approximately 63 million hectares of land are currently allocated for biofuels production worldwide.[5] By 2040, global biofuel production may double[6] or triple,[7] which could take up an additional 44 to 118 million hectares of land. This level of expansion would exceed a land area equivalent to the US state of Texas. The way in which this future land conversion occurs will impact biodiversity, ecosystems, and climate.[8] Land-use planning strategies are needed to meet energy demands while sustaining ecosystem services like species biodiversity and water quality.

Biofuels Impact Assessment: Making It Spatial and Marginal

The expansion of biofuels poses trade-offs with other land uses like food production or conservation, and can result in economic, environmental, and social consequences.[9] However, these trade-offs can be reduced when human activities are carefully located across a region, and lands are designated to crops and other areas set aside for conservation in a way that benefits both biofuels production and ecosystem services.[10]

Life-cycle assessments (LCAs) are commonly used to evaluate the sustainability of biofuels production.[11] LCAs determine a product's environmental impacts by considering its entire life cycle, from raw materials extraction through production, usage, and disposal. Land-use impacts measured by LCAs tend not to be regionally specific because they often aggregate information across different climatic regions, vegetation types, and biophysical conditions (e.g., soil types, geology).[12] As such, they fail to factor the significant influence of local landscape pattern on biodiversity and ecosystem processes. This approach also does not account for the marginal effect of biofuels or the impact of the next hectare of cropland or gallon of oil produced.[13] The marginal benefits or losses from the next unit of biofuel production can vary spatially across a landscape and are highly dependent on the type, amount, and distribution of habitats where land conversion occurs.[14] Measuring marginal changes is more relevant for land-use decision making that involves small-scale, iterative changes—for example, whether to convert a given land parcel for development.[15] LCA models also often assume a linear relationship between area impacted and the environmental response (e.g., species loss); thus potential nonlinearities, tipping points, and critical thresholds to land conversion will go undetected although they can determine the resilience of landscapes.[16] In our case study, we improve upon the LCA by conducting spatial analyses that account for the regional influence of landscape pattern, the marginal impact of biofuels, and the nonlinear relationships between biofuels expansion and provision of ecosystem services.

Testing Ground: Case Study in the Brazilian Cerrado

In 2012, The Nature Conservancy collaborated with The Dow Chemical Company to investigate the potential for targeted land-use planning to achieve both commodity production and environmental goals in the Brazilian Cerrado. The Cerrado is the world's most diverse tropical savanna, but more than half has been destroyed and continues to be threatened by biofuels expansion, along with other land uses.[17] Using the Dow subsidiary Santa Vitória Açúcar e Álcool (SVAA) as a case study, we focused on a watershed in Minas Gerais State that is largely pasture, a portion of which SVAA plans to convert to sugarcane fields for the production of ethanol (fig. 8-1a). In the study area, less than 20 percent of natural habitat remains and is made up of four dominant vegetation types (cerrado, cerradão, semideciduous forests, and wetlands). Twenty-five percent of leased farmland will be maintained in natural vegetation to comply with the Brazilian Forest Code,[18] a federal policy that promotes biodiversity and hydrologic services on private lands.[19] We examined how commercial sugarcane production can expand in a way that benefits both biodiversity and freshwater quality under existing environmental legislation in Brazil.

Balancing Nature and Economic Trade-Offs

We applied spatial optimization techniques to map marginal service values, assess economic and environmental trade-offs, and find efficient land-use patterns across the study watershed. Our analyses included

- Models that optimized agricultural profit (measured by sugarcane production and cattle ranching); biodiversity (measured by bird and mammal species); and water quality (measured by levels of nitrogen, phosphorous, and sediment retention).

- Maps of those values across the watershed at a small-scale that depicted where the next unit of habitat clearing (or restoration/protection) would be most costly or beneficial to profit and the two ecosystem services.

- Prioritization maps based on the order of land unit conversion that incrementally revealed the most important places for the different services across the watershed. The best 25 percent of the watershed was then highlighted to coincide with the Brazilian Forest Code's habitat threshold (figs. 8-1b–d).

- Efficiency frontiers that plotted agricultural profit against biodiversity or water quality to assess the marginal loss of those services per unit gained in profit and to detect potential land-use thresholds. We developed different scenarios in which land use was optimized for profit and water quality only, profit and biodiversity only, and profit and both biodiversity and water quality considered together (figs. 8-2a–b).[20]

Conflicts between Profit, Biodiversity, and Water Quality

We found widely varying profit, biodiversity, and water quality patterns across the region (figs. 8-1b–d).* The northern part of the watershed was most profitable to develop as sugarcane fields for ethanol, owing to the proximity of the sugarcane processing mill (fig. 8-1b). In contrast, larger, more connected patches of cerradão and semideciduous forest had the highest values for biodiversity because they provided the most suitable habitats for birds and mammals. Finally, wetlands and riparian forests along water bodies and on steep slopes had the highest values for prioritizing water quality because of their high filtering potential and / or retention of pollutants.

When considering the best 25 percent of lands for the different services, only 12 percent of the watershed exhibited spatial conflicts between making a profit and protecting the environment. The remaining 13 percent of lands with high profit potential had no spatial conflict with

---

* There was little overlap in the geographic distributions of sugarcane profit and biodiversity (Spearman's rank, $\rho = 0.47$) or water quality ($\rho = -0.05$), and between biodiversity and water quality ($\rho = -0.14$).

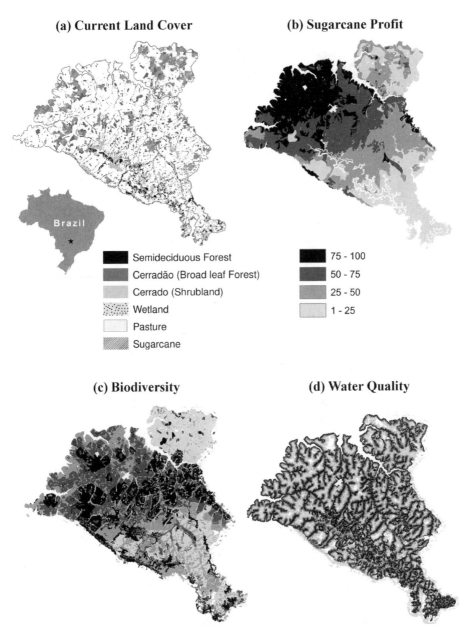

(a) Current Land Cover

(b) Sugarcane Profit

Brazil

Semideciduous Forest
Cerradão (Broad leaf Forest)
Cerrado (Shrubland)
Wetland
Pasture
Sugarcane

75 - 100
50 - 75
25 - 50
1 - 25

(c) Biodiversity

(d) Water Quality

Figure 8-1. (a) Current land cover and land use for the Ribeirão São Jerônimo study watershed in southeastern Brazil. Prioritization maps depicting the relativized marginal values for (b) sugarcane profit, (c) biodiversity, and (d) water quality in the watershed.

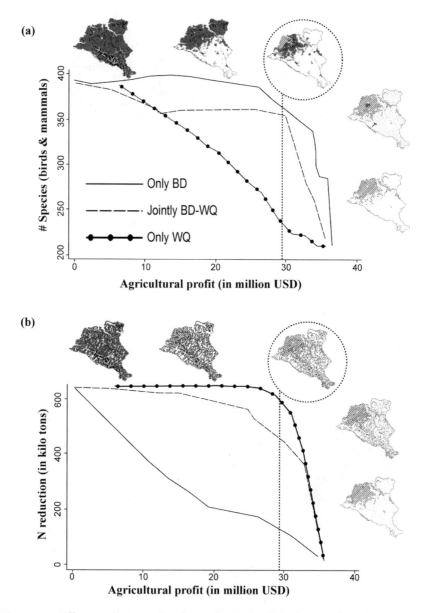

Figure 8-2. Efficiency frontiers that depict the trade-offs between agricultural profit and only biodiversity (Only BD); profit and only water quality (Only WQ); and profit and both BD and WQ (joint BD-WQ).

biodiversity or water quality and are expected to produce 1 million tons of sugarcane over the current annual sugarcane production target.

Not only was there limited overlap in areas best for profit and ecosystem services, but also in those best for water quality and biodiversity. Only 3 percent of the watershed fell in the top 25 percent for both of these ecosystem services (figs. 8-1c–d). This counters a commonly held assumption that optimizing for one conservation goal will benefit another. Indeed, lower biodiversity was predicted when planning prioritized only water quality and vice versa. Specifically, when Forest Code compliance maximized only water quality, 145 fewer species (69 percent) were predicted to remain in the landscape relative to the similar scenario maximizing biodiversity (figs. 8-2a–b). Similarly, when Forest Code compliance maximized only biodiversity, an additional 441 tons of nitrogen, 22 tons of phosphorous, and 67 tons of sediment were predicted to end up in waterways, reducing average water quality by almost 30 percent.

Multiservice Land Use Planning to Reduce Trade-Offs

By jointly optimizing biodiversity and water quality, our models predicted substantial environmental improvement. Specifically, 92 percent of species conservation, 73 percent of nitrogen and phosphorous, and 94 percent of sediment retention was possible under joint optimization relative to when only one ecosystem service was maximized (figs. 8-2a–b). Additionally, compared with the current landscape, this multiservice planning resulted in a predicted sixty-nine additional species and a 13 percent improvement in average water quality while still increasing agricultural profit by $6.5 million (2016 U.S. dollars).

Multiservice Land Use Planning to Find Tipping Points

We also found that gains in profit could only be obtained at high costs to ecosystem services when agriculture spanned more than 75 percent of the watershed. Conversely, costs to both biodiversity and water quality remained low if agriculture converted less than 75 percent of the natural habitat within the landscape while optimizing for ecosystem services. Notably, when agriculture expanded without spatially targeting environmental benefits, our models predicted a continual, linear loss in both biodiversity and water quality. The presence of pronounced thresholds and the potential for steep declines in both biodiversity and water quality raise questions about the effectiveness of the Forest Code in the absence of full and optimal compliance, which is the norm in this region and elsewhere in Brazil.[21]

Sustainable Landscape Design: Moving from Potential to Practice

Our case study demonstrates that both biofuels production and environmental goals can be met with landscape planning, even in a biodiversity and agricultural hotspot like the Brazilian Cerrado. We show that potential gains in profit and the resulting losses of ecosystem services vary spatially and can exhibit nonlinear relationships that result in critical land-use thresholds. Encouragingly, our results indicate that sugarcane production can expand with little or no environmental detriment if land-use planners take into account spatial heterogeneities and consider multiple ecosystem services at a large scale (in this case, across an entire watershed).

Spatial optimization tools, like ours, are increasingly available to model land-use impacts, estimate ecosystem changes, and tailor biofuels assessments to local conditions to inform the design of sustainable landscapes.[22] Implementing such landscape designs, however, will require mechanisms that promote the adoption of multiservice landscape planning and foster coordination across properties. One regulatory mechanism that countries rely on to avoid, minimize, and compensate (offset) for development impacts is impact mitigation;[23] policies are emerging that incentivize the adoption of landscape principles into mitigation planning and practice.[24] However, agricultural activities are often exempt from mitigation licensing, except in countries like Brazil and Australia that have land-clearing laws that regulate the removal of native vegetation.[25] Governments can also impose land-use requirements to influence the zoning of legally convertible areas for biofuels (e.g., degraded lands) along with areas off-limits to development. Such land-use zoning approaches have been adopted in Brazil for sugarcane[26] and in Indonesia for oil palm.[27]

Other mechanisms include voluntary commitments—for example sustainability certification schemes and zero-deforestation commitments—and payments for ecosystem services—for example, biodiversity banking, water funds, and REDD ("Reducing Emissions from Deforestation and Degradation") carbon projects.[28] Industry stakeholder forums, such as the Roundtables for Responsible (or Sustainable) Soy, Sugar, Palm Oil, and Biofuels, and commodity certification schemes have standards aimed at shifting business practices to locate and carry out production in a way that minimizes habitat clearing, maintains ecosystem services, and protects high-conservation-value areas.[29] A more systematic incorporation of landscape design criteria into sustainability standards, high-conservation-value-area designation, ecocertification, and payment-for-ecosystem-services schemes is recognized as needed to broaden their site-based focus.[30] Collectively, such legal, voluntary, and market mechanisms, if effectively implemented and widely adopted by agricultural producers in affected landscapes, hold promise of delivering large-scale conservation outcomes in the expanding biofuels sector.

Figure 8-3. *Sugarcane fields in the Cerrado landscape in Minas Gerais State, in Brazil. Photo credit: Jennifer Molnar.*

Acknowledgments

We are grateful to E. Okumura, E. Garcia, C. Pereira, and J. Pereira for input on the sugarcane modeling; to L. Azevedo, J. Guimarães, K. Voss, I. Alameddine, S. Thompson, and B. Keeler for input on the hydrologic modeling; and to E. Lonsdorf and E. Nelson for input on the biodiversity and optimization modeling. Funding was provided by The Dow Chemical Company Foundation, The Dow Chemical Company, The Nature Conservancy, The Anne Ray Charitable Trust, and The 3M Foundation.

Notes

1. José Escobar, Electo Lora, Osvaldo Venturini, Edgar Yáñez, Edgar Castillo, and Oscar Almazan, "Biofuels: Environment, Technology and Food Security," *Renewable and Sustainable Energy Reviews* 13 (2009): 1275–87, doi:10.1016/j.rser.2008.08.014.
2. Ibid.
3. IEA, *World Energy Outlook 2015* (see ch. 1, n. 3).
4. International Energy Agency, *Technology Roadmap: Biofuels for Transport*, 2011, http://www.iea.org/publications/freepublications/publication/biofuels_roadmap_web.pdf.
5. Ronald Conte, "How Much Farmland Is Used for Biofuel?," *Hunger Math: World Hunger by the Numbers* (blog), Oct. 29, 2015, https://hungermath.wordpress.com/2015/10/29/how-much-farmland-is-used-for-biofuel/.
6. US IEA, *Annual Energy Outlook 2013* (see ch. 4, n. 15).
7. US IEA, *World Energy Outlook 2015* (see ch. 1, n. 3).
8. Escobar et al., "Biofuels: Environment, Technology and Food Security"; Joseph Fargione, Richard Plevin, and Jason Hill, "The Ecological Impact of Biofuels," *Annual Review of Ecology, Evolution, and Systematics,* 41 (2010): 351–77, doi:10.1146/annurev-ecolsys-102209-144720.
9. Ibid.
10. Lian Pin Koh and Jaboury Ghzoul, "Spatially Explicit Scenario Analysis for Reconciling Agricultural Expansion, Forest Protection, and Carbon Conservation in Indonesia," Proceedings of the National Academy of Sciences of the United States of America 107, no. 24 (2010): 11140–11144, doi: 10.1073/pnas.1000530107
11. Fargione et al., "Ecological Impact of Biofuels."
12. Michael Curran, Laura de Baan, An De Schryver, Rosalie Van Zelm, Stefanie Hellweg, Thomas Koellner, Guido Sonnemann, and Mark Huijbregts, "Toward Meaningful End Points of Biodiversity in Life Cycle Assessment," *Environmental Science Technology* 45 (2010): 70–79, doi:10.1021/es101444k.
13. Fargione et al., "Ecological Impact of Biofuels."
14. Taylor Ricketts and Eric Lonsdorf, "Mapping the Margin: Comparing Marginal Values of Tropical Forest Remnants for Pollination Services," *Ecological Applications* 23, no. 5 (2013): 1113–23, doi:10.1890/12-1600.1.
15. Ibid.
16. Curran et al., "Toward Meaningful End Points."

17. David Lapola, Ruediger Schaldach, Joseph Alcamo, Alberte Bondeau, Jennifer Koch, Christina Koelking, and Joerg Priess, "Indirect Land-Use Changes Can Overcome Carbon Savings from Biofuels in Brazil," *Proceedings of the National Academy of Sciences of the United States of America* 107, no. 8 (2010): 3388–93, doi:10.1073/pnas.0907318107.

18. Christina Kennedy, Daniela Miteva, Leandro Baumgarten, Peter Hawthorne, Kei Sochi, Stephen Polasky, James Oakleaf, Elizabeth Uhlhorn, and Joseph Kiesecker, "Bigger Is Better: Improved Nature Conservation and Economic Returns from Landscape-Level Mitigation," *Science Advances* 2, no. 7 (2016):e1501021, doi:10.1126/sciadv.1501021.

19. Britaldo Soares-Filho, Raoni Rajão, Marcia Macedo, Arnaldo Carneiro, William Costa, Michael Coe, Hermann Rodrigues, and Ane Alencar, "Cracking Brazil's Forest Code," *Science* 344, no. 6182 (2014): 363–64, doi:10.1126/science.1246663.

20. Christina Kennedy, Peter Hawthorne, Daniela Miteva, Leandro Baumgarten, Kei Sochi, Marcelo Matsumoto, Jeffrey Evans, Stephen Polasky, Perrine Hamel, Emerson Vieira, Pedro Ferreira Develey, Cagan Sekercioglu, Ana Davidson, Elizabeth Uhlhorn, and Joseph Kiesecker, "Optimizing Land Use Decision-Making to Sustain Brazilian Agricultural Profits, Biodiversity, and Ecosystem Services," *Biological Conservation* 204 Part B 2016:221–230, doi: 10.1016/j.biocon.2016.10.039.

21. Soares-Filho et al., "Cracking Brazil's Forest Code."

22. Patrick O'Farrell and Pippin Anderson, "Sustainable Multifunctional Landscapes: A Review to Implementation," *Current Opinion in Environmental Sustainability* 2, no. 1–2 (2010): 59–65, doi:10.1016/j.cosust.2010.02.005.

23. Becca Madsen, Nathaniel Carroll, Daniel Kandy, and Genevieve Bennett, "2011 Update: State of Biodiversity Markets" (Washington, DC: Forest Trends, 2011), http://www.ecosystemmarketplace.com/reports/2011_update_sbdm.

24. David Hayes, "Addressing the Environmental Impacts of Large Infrastructure Projects: Making 'Mitigation' Matter," *Environmental Law Reporter* 44 (2014): 10016–21.

25. Madsen et al., "2011 Update."

26. IEA, *Technology Roadmap.*

27. Betsy Yaap, Matthew Struebig, Gary Paoli, and Lian Koh, "Mitigating the Biodiversity Impacts of Oil Palm Development," *CAB Reviews: Perspectives in Agriculture, Veterinary Science, Nutrition and Natural Resources* 5, no. 019 (2010): 1–11.

28. Yaap et al., "Mitigating the Biodiversity Impacts."; Rebecca Chaplin-Kramer, Malin Jonell, Anne Guerry, Eric Lambin, Alexis Morgan, Derric Pennington, Nathan Smith, Jane Atkins Franch, and Stephen Polasky, "Ecosystem Service Information to Benefit Sustainability Standards for Commodity Supply Chains," *Annals of the New York Academy of Sciences* 1355 (2015): 77–97, doi:10.1111/nyas.12961.

29. Ibid.

30. Lian Koh, Patrice Levang, and Jaboury Ghazoul, "Designer Landscapes for Sustainable Biofuels," *Trends in Ecology and Evolution* 24, no. 8 (2009): 431–38, doi:10.1016/j.tree.2009.03.012.; Jaboury Ghazoul, Claude Garcia, and C.G. Kushalappa, "Landscape Labelling: A Concept for Next-Generation Payment for Ecosystem Service Schemes," *Forest Ecology and Management* 258, no 9 (2009): 1889–95, doi:10.1016/j.foreco.2009.01.038; Teja Tscharntke, Jeffrey Milder, Götz Schroth, Yann Clough, Fabrice DeClerck, Anthony Waldron, Robert Rice, and Jaboury Ghazoul, "Conserving Biodiversity through Certification of Tropical Agroforestry Crops at Local and Landscape Scales," *Conservation Letters* 8, no. 1 (2015): 14–23, doi:10.1111/conl.12110.

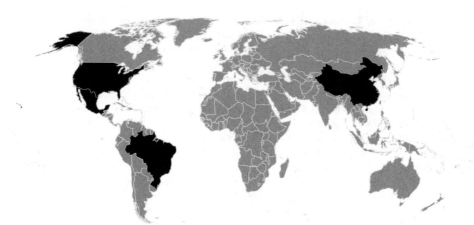

CHAPTER 9

Sustainable Energy and Healthy Rivers

Jeff J. Opperman

W e know that the world must develop energy systems that support healthy, prosperous lives for people, allow the world to remain within safe climate boundaries, *and* accomplish those goals without causing unacceptable impacts on ecosystems and their services, vulnerable communities, and irreplaceable natural values. Hydropower is a clear illustration of this central challenge and opportunity.

Electricity generated from hydroelectric dams is currently the world's largest, most mature, and most reliable source of low-carbon, renewable energy. (The Intergovernmental Panel on Climate Change estimates that life-cycle emissions from hydropower are 5 percent that of natural gas and three percent that of coal,[1] though note that certain types of reservoirs, particularly in the tropics, can have significant emissions and reservoir emissions are a topic of ongoing research.[2]) Global energy projections, particularly those that assume policies to promote transition from fossil fuels, forecast an approximate doubling of global hydropower capacity by 2050.[3] Achieving this doubling could result in widespread negative impacts to river ecosystems and freshwater species, and serious economic and social disruption for the tens of millions to hundreds of millions of people who depend on functioning river systems for food and livelihoods. Social and environmental losses at this scale would arguably negate the climate and energy benefits of hydropower expansion. From the perspective of the energy sector, high levels of social and environmental opposition increase uncertainty and constrain development and/or operation of projects.

This chapter explores two related premises. First, that hydropower can both play an important role in future energy systems and contribute to climate goals. Second, that new and existing hydropower projects must improve significantly in social and environmental performance to avoid unacceptable environmental and social impacts. In fact, without improved performance that seeks to reduce conflicts, hydropower may not reach its potential contribution to sustainable energy systems.

The Big Picture

As of 2012, renewable sources of electricity generation produced 4,800 terawatt hours (TWh), representing 21 percent of total global annual electricity generation. Of that renewable total, hydropower provided 3,670 TWh, or just over 75 percent and approximately six times more than wind and solar combined.[4] While wind and solar have made significant strides in recent years, most projections for future renewable energy scenarios still include a major role for hydropower. For example, the 2040 projection of the International Energy Agency (IEA) for meeting climate objectives envisions renewable energy generation reaching 17,970 TWh (51 percent of global generation), with much of that increase coming from a tenfold increase in wind and a thirtyfold increase in solar. Even though this projection forecasts a much lower relative increase for hydropower, it still has hydropower as the largest single source of renewable energy, providing 40 percent of the total renewable energy generation—nearly a doubling of current generation to 6,940 TWh.

Hydropower also provides a particularly valuable set of energy services, including both storage for baseload generation and the ability to quickly respond to fluctuations in demand. Hydropower reservoirs that can store water are in effect storing energy. Thus one of hydropower's key contributions to a sustainable energy future is that its ability to store energy can catalyze greater expansion of intermittent renewables, such as wind and solar.

However, all of these energy services come with the potential for large environmental and social impacts.[5] Hydropower dams and their reservoirs have a large footprint on the landscape, including the displacement of communities and the inundation of agricultural land or specific ecosystems. In 2000, the World Commission on Dams estimated that 40 to 80 million people had been displaced by dams in the past century. In addition to the direct footprint, the environmental and social impacts of dams extend both upstream and downstream, as when dams block the movement of fish and other migratory aquatic species that often have high biological, cultural, and economic value.

Further, changes to the flow regime also can affect ecosystems and processes that maintain important services for people, such as river-floodplain connectivity important for fisheries and agriculture.[6] Scientists estimate that the number of people affected by downstream impacts may be several times larger than the number affected by direct displacement.[7] Water management infrastructure, such as dams, consistently ranks among the leading causes of decline of freshwater species.[8]

The projected expansion of hydropower dams will affect more than 300,000 kilometers of river channel globally by fragmentation, flow alteration, or both. Nearly 70 percent of this loss of free-flowing river channel will occur in freshwater ecoregions that harbor the greatest diversity of fish species. Rivers that currently feed millions of people—including the Mekong, Irawaddy, Amazon, and Ganges—are projected to be among the most impacted river basins.[9]

Examples of a Way Forward

There is a way forward. Both older and more recent innovations highlight a path in which hydropower contributes to a sustainable energy future. On this path, existing dams improve their environmental and social performance, while new dams are planned and sited in a way that is consistent with healthy rivers and local communities. Further, hydropower is managed and planned in a way that facilitates expansion of renewables—maximizing its mitigation benefits for climate change—and enhances other water-management objectives (water supply, flood-risk management, ecosystem restoration) that contribute to adaptation to climate change.

This optimistic vision hinges on moving away from approaching hydropower management and planning at the project scale and instead looking at the entire system. However, most governments conduct reactive environmental reviews in response to developers' proposals for new licenses or relicensing of existing dams, rather than strategic planning and management for the entire system. A system-scale approach allows government planners and regulators to explore, compare, and choose the best alternative to meet multiple objectives. The following section illustrates hydropower planning and management at system scales in three brief case studies.

Three Gorges Dam, China. Individual existing hydropower projects can be "reoperated" to improve environmental performance. For example, several years ago, the Three Gorges Dam on the Yangtze River in China began releasing water in the late spring to promote spawning conditions for native carp species (an environmental flow). However, implementation of environmental flows lags far behind the science, largely owing to conflicts from other water-management objectives (e.g., environmental flow releases reducing electricity generation). System-scale approaches, in which environmental flows are incorporated into processes to reoptimize an overall water management system, may identify solutions that improve environmental performance while maintaining or even enhancing energy generation.[10]

Penobscot River, Maine. Decades of contentious licensing or relicensing of individual dams failed to resolve conflicts between energy generation and migratory fish passage on the Penobscot River in the state of Maine. In the early 2000s, a single hydropower company acquired the major dams on the Penobscot mainstem. They worked with the Penobscot

Indian Nation, state and federal agencies, and conservation organizations on an agreement that removed two dams and bypassed a third with a nature-like fish passage. Owing to operational and equipment changes at the remaining dams on the Penobscot, the total energy generated within the Penobscot will remain the same, or increase slightly.[11] Biologists believe that habitat for migratory fish will dramatically improve by removing the dams. Initial results indicate that populations of river herring have increased by several orders of magnitude. Moving toward the system scale allowed for a broader range of solutions and balanced outcomes than could be achieved at the scale of individual projects.

Coatzacoalcos River, Mexico, and Tapajos River, Brazil. Modeling in these two river basins illustrates the potential for system-scale planning—integrating hydropower with protection of environmental and social resources—to identify options that balance energy development and conservation of other values. In Mexico, The Nature Conservancy collaborated with the Federal Commission for Electricity and the National Commission for the Knowledge and Use of Biodiversity to simulate twenty-five hydropower development scenarios in the Coatzacoalcos River basin, using an inventory of potential dams from the federal commission, and modeled the performance of those scenarios across a range of environmental and social resources (fig. 9-1). Impacts of scenarios varied widely; for example, while some scenarios resulted in significant fragmentation and flow alteration across the basin, other scenarios could achieve a majority of the basin's energy potential with relatively limited impacts on the extent of free-flowing rivers.

A similar modeling approach for the Tapajos River basin (a tributary to the Amazon in Brazil) also illustrated the potential for system-scale planning to identify more balanced outcomes for hydropower development. For example, comparing two options that achieved approximately 60 percent of the basin's hydropower capacity, a scenario selecting a mix of projects based on least cost (typical criteria of economic feasibility) would fragment 5,150 kilometers of river channel, while a scenario that balances generation with river connectivity would fragment 3,200 kilometers. The second scenario would produce the same energy, for slightly higher costs (5 percent) but leave 2,000 additional kilometers of free-flowing river.

From Best Practice to Common Practice

The licensing and relicensing process of hydropower in the United States is considered one of the better examples of a regulatory approach that seeks to periodically revisit decisions and revise management to achieve better balance between energy, environmental and social performance. The Penobscot example shows that there is flexibility within the current regulatory approach to move from a project focus to system-scale solutions. It will be challenging, however, to find solutions in river basins with multiple ownerships rather than a single owner. Mitigation banking, currently used with a range of environmental regulatory contexts, could be adapted to allow hydropower licensing, or other regulatory processes, to promote system-scale environmental benefits when meeting mitigation requirements.[12] Analysis at the system scale can often identify mitigation opportunities with disproportionately large benefits. For example, removing twelve dams in the Willamette Basin (Oregon) could reconnect more than half of the drainage network, greatly increasing habitat for migratory fish such as salmon, while losing less than 2 percent of the basin's hydropower capacity.[13] A mitigation banking system could allow pooled funding to be directed toward those few dams whose removal would have disproportionately large gains for river connectivity, linking management of individual projects toward system-scale benefits.

Most countries undergoing rapid hydropower expansion are unlikely to prioritize environmental protection in the short term. Even countries in the Mekong River Basin, which has one of the most valuable freshwater fisheries feeding tens of millions of people, are rapidly building projects that will greatly diminish that environmental value of free-flowing rivers.[14] However, system-scale approaches to planning and management can help solve numerous other challenges associated with hydropower development; improved environmental and social performance may be a secondary, though significant, benefit.

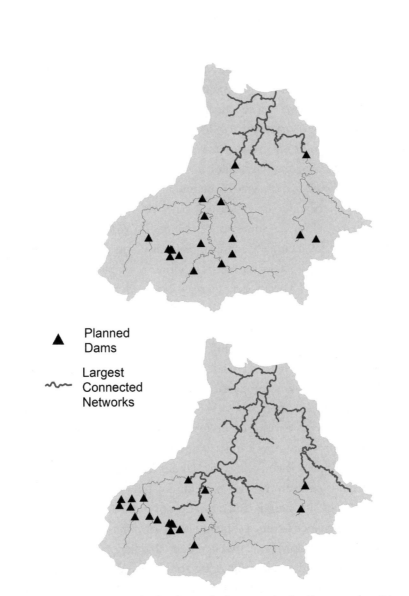

Figure 9-1. *Two scenarios for developing hydropower in the Coatzacoalcos River basin that provide similar energy (300 MW, or approximately 60 percent of total basin capacity), but with considerably different lengths of channel affected by fragmentation: The scenario on the top has nearly 1,000 kilometers impacted compared with the scenario on the bottom, which shows approximately 400 km impacted (data from CFE and CONABIO). The Coatzacoalcos is located in southeast Mexico and flows northward to the Gulf of Mexico.*

First, hydropower projects often suffer from delays and cost over-runs,[15] and conflict over environmental and social impacts contributes to these challenges. In several recent high-profile examples, concerns over environmental and social impacts led to suspension or cancellation of projects after significant investments had been made (e.g., Myitsone in Myanmar and HidroAysen in Chile). System planning approaches, particularly if conducted in a transparent manner, could identify potential risks and help planners and managers avoid major conflicts. In the 1980s, Norway conducted a national survey of remaining hydropower potential and, using criteria that included technical feasibility and environmental and social values, categorized river basins as either available for development or to be protected.[16] This process increased certainty for both developers and conservationists, reduced conflicts during subsequent development,[17] and, where replicated, may help facilitate financing.

Second, planning and developing a sequence of single projects is much less likely to produce well-integrated infrastructure systems for energy and water management compared with system planning. In some situations, single-project approaches even result in new dams that interfere with the operation of existing dams. Optimization methods can be used to inform decisions about how infrastructure investments can work together to achieve multiple societal objectives. Under this approach, models compare thousands or even millions of different infrastructure portfolios—permutations of site, design, and operation of individual dams—in terms of their performance across a set of metrics for different values (e.g., hydropower, water supply, flood management, and fisheries). The modeling results quantify and visually illustrate trade-offs, allowing stakeholders and decision makers to understand their options.[18] Ultimately, this approach allows governments to identify a set of infrastructure investments that can deliver the broadest benefits to their people.[19]

This multiobjective optimization approach can identify low-carbon energy systems that meet economic needs while maintaining or restoring other environmental and social values. To meet this challenge, hydropower shouldn't be planned or managed in isolation, even if at large geographic scales. Instead, hydropower decisions should be made within

Figure 9-2. *The massive Itaipu hydroelectric dam (second in size only to China's Three Gorges) located on the Paraná River bordering Brazil and Paraguay. Construction of the dam started in 1975, and the dam started generating electricity in 1984. Photo credit: Scott Warren.*

Figure 9-3. *Three Gorges Dam. The Yangtze River flows across China and empties into the East China Sea near the historic city of Shanghai. The river has great cultural significance and has provided food and livelihoods for millions of people who have lived along its shores for centuries. This river and others have become a source of carbon-neutral electricity for China but by working with the hydropower industry— including the China Three Gorges Corporation—these projects can be better planned, designed, and operated. Photo credit: Kevin Arnold.*

Figure 9-4. *The Itaipu hydroelectric dam is 196 meters (643 ft.) high, equivalent to a sixty-five-story building. For the 14,000 MW installed power, 1,350 square kilometres (520 sq mi) were flooded. While this footprint is large, if Brazil were to use standard power generation to produce the electric power of Itaipu, 434,000 barrels of petroleum would have to be burned every day. Photo credit: Erika Nortemann/The Nature Conservancy.*

the context of broader goals for emissions and energy, including other renewables, and the context of environmental and social values from rivers and other ecosystems. These three great challenges—energy, climate, and environmental protection—cannot be addressed in isolation, and the most promising solutions are likely to be found through integration.

Acknowledgments

Thanks to David Harrison, Joerg Hartmann, and Justus Raepple for constructive review of this chapter.

Notes

1. Thomas Bruckner, Lew Fulton, Edgar Hertwich, Alan McKinnon, Daniel Perczyk, Joyashree Roy, Roberto Schaeffer, Steffen Schlömer, Ralph Sims, Pete Smith, and Ryan Wiser (Steffen Schlömer, ed.), "Annex III: Technology-specific cost and performance parameters," in *Climate Change 2014: Mitigation of Climate Change. Contribution of Working Group III to the Fifth Assessment Report of the Intergovernmental Panel on Climate Change* (2014).

2. Bridget Deemer, John Harrison, Siyue Li, Jake Beaulieu, Tonya DelSontro, Nathan Barros, José Bezerra-Neto, Stephen Powers, Marco dos Santos, and J. Arie Vonk, "Greenhouse Gas Emissions from Reservoir Water Surfaces: A New Global Synthesis," *BioScience* 66, no. 11 (2016): 949–64.

3. International Energy Agency (IEA). (2014). World energy outlook 2014. Retrieved from http://www.worldenergyoutlook.org/publications/weo-2014.

4. Ibid.

5. World Commission on Dams, *Dams and Development: A New Framework for Decision-Making: The Report of the World Commission on Dams* (Earthscan, 2000); Franklin Ligon, William Dietrich, and William Trush, "Downstream Ecological Effects of Dams," *Bio-Science* 45, no. 3 (1995):183–92.

6. Jeffrey Opperman, Gerald Galloway, and Stéphanie Duvail, "The Multiple Benefits of River-Floodplain Connectivity for People and Biodiversity," in *Encyclopedia of Biodiversity* 2nd ed., no. 7 (Waltham, MA: Academic Press, 2013),144–60.

7. Brian Richter, Sandra Postel, Carmen Revenga, Thayer Scudder, Bernhard Lehner, Allegra Churchill, and Morgan Chow, "Lost in Development's Shadow: The Downstream Human Consequences of Dams," *Water Alternatives* 3, no. 2 (2010).

8. Brian Richter, David Braun, Michael Mendelson, and Lawrence Master, "Threats to Imperiled Freshwater Fauna," *Conservation Biology* 11, no. 5 (1997): 1081–93, doi:10.1046/j.1523–1739.1997.96236.x; Robert McDonald, Julian Olden, Jeffrey Opperman, William Miller, Joseph Fargione, Carmen Revenga, Jonathan Higgins, and Jimmie Powell, "Energy, Water and Fish: Biodiversity Impacts of Energy-Sector Water Demand in the United States Depend on Efficiency and Policy Measures," *PloS One* 7 no. 11 (2012), doi:10.1371/journal.pone.0050219.

9. Jeffrey Opperman, Gunther Grill, and Joerg Hartmann, "The Power of Rivers: Finding

Balance between Energy and Conservation in Hydropower Development," *The Nature Conservancy* (2015); Kirk Winemiller, Peter McIntyre, Leandro Castello, Etienne Fluet-Chouinard, et al., "Balancing Hydropower and Biodiversity in the Amazon, Congo, and Mekong," *Science* 351, no. 6269 (2016): 128–29, doi:10.1126/science.aac7082.

10. Se-Yeun Lee, Carolyn Fitzgerald, Alan Hamlet, and Stephen Burges, "Daily Time-Step Refinement of Optimized Flood Control Rule Curves for a Global Warming Scenario," *Journal of Water Resources Planning and Management* 137, no. 4 (2010): 309–17.

11. Jeffrey Opperman, Joshua Royte, John Banks, Laura Rose Day, and Colin Apse, "The Penobscot River, Maine, USA: A Basin-Scale Approach to Balancing Power Generation and Ecosystem Restoration," *Ecology and Society* 16, no. 3 (2011): 7.

12. Dave Owen and Colin Apse, "Trading Dams," *UC Davis Law Review* 48 (2014): 1043.

13. Michael Kuby, William Fagan, Charles ReVelle, and William Graf, "A Multiobjective Optimization Model for Dam Removal: An Example Trading Off Salmon Passage with Hydropower and Water Storage in the Willamette Basin," *Advances in Water Resources* 28, no. 8 (2005): 845–55.

14. Guy Ziv, Eric Baran, So Nam, Ignacio Rodríguez-Iturbe, and Simon A Levin, "Trading-Off Fish Biodiversity, Food Security, and Hydropower in the Mekong River Basin," *Proceedings of the National Academy of Sciences* 109, no. 15 (2012):5609–14.

15. Atif Ansar, Bent Flyvbjerg, Alexander Budzier, and Daniel Lunn, "Should We Build More Large Dams? The Actual Costs of Hydropower Megaproject Development," *Energy Policy* 69 (2014):43–56.

16. Svein Tore Halvorsen, "The Master Plan for the Management of Watercourses in Norway," *Environmentalist* 8, no. 1 (1988): 39–45.

17. Sigmund Huse, "The Norwegian River Protection Scheme: A Remarkable Achievement of Environmental Conservation," *Ambio* 16, no. 5 (1987): 304–8.

18. Anthony Hurford, Ivana Huskova, and Julien J. Harou, "Using Many-Objective Trade-Off Analysis to Help Dams Promote Economic Development, Protect the Poor and Enhance Ecological Health," *Environmental Science & Policy* 38 (2014): 72–86; Anthony Hurford and Julien Harou, "Balancing Ecosystem Services with Energy and Food Security—Assessing Trade-Offs for Reservoir Operation and Irrigation Investment in Kenya's Tana Basin," *Hydrology and Earth System Sciences* 11, no. 1 (2014): 1343–88.

19. The Nature Conservancy, WWF, and the University of Manchester, "Improving Hydropower Outcomes through System-Scale Planning: An Example From Myanmar," a report for the United Kingdom's Department for International Development (The Nature Conservancy, 2016).

PART III

Making Best Practice Common Practice

Principles of sustainability, although proven in concept, have not yet reached the public consciousness. Success stories from part II will not become commonplace until conveyed to decision makers, adopted by industry, and implemented into policy. Part III builds the case to more rapidly embrace comprehensive energy planning. Here we identify next steps that the energy industry, the environmental community, the banking sector, and governments can take to build sustainability into our energy future.

Industry is a major force for change, and companies are increasingly adopting sustainability principles into their cost of doing business as governments shift from voluntary to mandatory statutes. Nature gives most industries the basic building blocks for their product line and ultimately their bottom line. As society adopts sustainability, so will companies, and they'll invest in conservation because it makes good business sense. Industry brings to conservation an unparalleled ability to move human, physical, and financial capital around the globe; ownership and management of extensive land and resource holdings in biodiversity-rich regions; access to supply chains that draw from and influence a wide array of natural resources; and corporate images that can influence consumer preferences and shape regional development patterns. Done right, industry contributions to conservation are an enormous and welcome investment that can help stem the loss of ecosystem services and associated biodiversity on global scales.

Industry will likely embrace the kinds of planning processes outlined in part II if doing so improves the certainty of their investments. Clearly

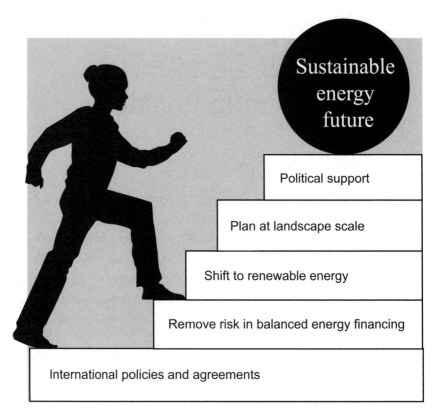

Figure p3-1. Steps that will advance comprehensive energy planning and lead to sustainable development.

governments and the environmental community need to meet energy developers halfway to help facilitate energy access and emission reductions goals by removing barriers that forestall development, in particular renewables development. Take, for example, the solar energy sector. Solar energy is taking off, but to win a race against carbon emissions, it will have to soar. The technology behind solar power continues to improve efficiency and drop in price. As it does so, the soft costs—permitting, financing, and installation—make up a bigger portion of the price tag for new projects. The US Department of Energy recently reported that soft costs now make up more than half the cost of installed solar power. Similar issues exist for wind and hydropower, the latter often delayed decades owing to permitting and financing issues. Cutting the red tape for renewables is a critical ingredient that rounds out a recipe for rapid renewable proliferation. The environmental community could play a key role in streamlining the licensing process by strategically identifying go-areas, as illustrated in chapters 4 and 5.

Access to capital, particularly in developing countries, also is critical, highlighting the need to work through other avenues, such as the International Finance Corporation, which sets standards for more than seventy Equator Principles financial institutions. Their current standards require project developers to follow principles of the mitigation hierarchy, including avoidance of impacts in critical habitat and achievement of net gains in biodiversity. Applying standards like these more widely will not be easy and will require political will, financial commitments, and improved capacity. However, these investments will pay dividends that include both development that's compatible with human well-being and reduced conflicts for both biodiversity conservation and society.

National-level governmental policy is an obvious mechanism for advancing smart energy planning. Already some seeds of change have been sown, and in recent years the thinking about this avenue has evolved. In the United States, for example, presidential executive order no. 13604 in 2012 and a presidential memorandum in 2015 provide for compensatory mitigation with mandatory adoption of a landscape-scale approach to facilitate investment in key regional conservation priorities and to ensure early integration of mitigation considerations in development projects. Other countries are following suit with at least forty-five mitigation programs for impact mitigation worldwide, and another twenty-seven in development. In 2012, for example, Colombia passed Resolution 1517 to require that the amount and location of compensation for development impacts are based on a series of landscape features. Despite progress, advancing mitigation remains a challenge, especially given the variability in national-level enabling conditions.

A number of global policy developments could provide the enabling conditions needed for national-level policy change that can advance comprehensive energy planning and combat energy sprawl. Of note is the

United Nations Climate Change summit in Paris in December 2015. The ensuing Paris Agreement serves as a foundation for all nations to limit global temperature rise to well below 2° Celsius (3.6°F), with an aspiration to reach 1.5° Celsius (2.7°F), and to adapt to climate change impacts already unfolding. With buy-in from nearly every country in the world, the agreement represents the single most important collective action for addressing climate change. The deal asks the 189 nations that signed it to reduce greenhouse gas emissions and requires that ratifying nations "peak" their greenhouse gas emissions as soon as possible and pursue the highest-possible emission reductions that each country can achieve.

The Paris Agreement is clearly a vital shift away from carbon-based energy sources, and it also affirms the important role that ecosystems and land use can play in reducing greenhouse gas emissions. It also promotes sustainable management of land, which can include conserving and restoring forests. Countries have agreed to publicly outline what climate actions they intend to take under a new international agreement, known as their Intended Nationally Determined Contributions. Most of these include the advancement of renewable energy, but many also include restoration and improved land management practices. Given the large land requirements of renewable energy, this could create the seemingly conflicting goals of advancing renewables while protecting and restoring natural habitat. In turn, the Intended Nationally Determined Contributions could be an impetus for national-level policy that requires large-scale land-use planning to balance competing outcomes.

There is no one silver bullet that will facilitate the widespread adoption of the practices outlined in part II. Getting to a sustainable energy future will certainly involve a herculean effort that engages industry, the environmental community, national governments, and international agencies. With insights from both the conservation community and development banks, chapter 10 distills the myriad of opportunities to advance comprehensive energy planning and highlights the key policies and practices that create pathways of least resistance to move the best practices highlighted in part II into common practice.

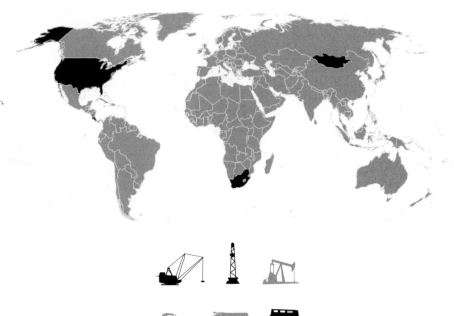

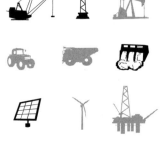

Policies, Practices, and Pathways for Sustainable Energy

Linda Krueger, Bruce McKenney,
Graham Watkins, and Amal-Lee Amin

In part I of this book, we provided an overview of the challenges for meeting future energy demand, demonstrating that regardless of the future energy mix, energy sprawl will be significant and likely lead to land-use conflict. In part II we highlighted best-practice options for reducing this conflict, offering innovative solutions to improve the future energy footprint. Now in this chapter we turn our attention to steps that establish those best practices as common practice for achieving sustainable energy landscapes that meet energy, economic, social, and environmental goals.

Future energy development will be shaped by the need to meet energy demand, increase energy access for 1.2 billion people with little or no access now, mitigate climate change, and close the $28 trillion gap in energy infrastructure financing (2016 US dollars). These drivers are central to four recent global policy developments that provide a platform for rethinking how to achieve sustainable energy landscapes:

- **Sustainable development**: Adoption of the United Nations Sustainable Development Goals as a development agenda from 2015 to 2030 set forth multiple goals and targets for energy access, infrastructure, the deployment of clean energy, and climate change that nearly all countries in the world have agreed to execute.

- **Climate change:** The United Nations Framework Convention on Climate Change Conference of the Parties in Paris in December 2015 established a legally binding agreement that includes emission reduction commitments from 189 countries. The Intended Nationally Determined Contributions submitted by parties imply significant changes to energy infrastructure to achieve these targets.

- **Energy access:** G20 leaders met in November 2015 and called for implementing the G20 Energy Access Action Plan to meet the needs of 1.2 billion people lacking access to electricity and 2.7 billion people who still rely on biomass for cooking.

- **Infrastructure finance:** The Addis Ababa Action Agenda established the Global Infrastructure Forum to address the infrastructure gap and invest in sustainable and resilient energy infrastructure through enhanced financial and technical support.

In this chapter we argue that the prominence of climate, energy access, and infrastructure finance on the global stage means that governments, international agencies, and the private sector are keen to work on developing sustainable energy solutions. We describe new policies and planning models that help accelerate deployment of climate-friendly renewables, reduce risks to investors, and improve the overall social and environmental performance of projects. We highlight examples of how national governments and international institutions can harness these solutions to move the global energy transition onto a sustainable path.

Shaping the Energy Future

Society needs a new road map for assessing synergies and trade-offs among multiple goals in energy development. Indeed, designing sustainable energy futures must be seen as a geographical process requiring integrated planning from national to local scales.[1] But this is not common practice. Demand, access, climate, and finance shape this road map to our energy future. The regional importance of each of these drivers depends on the country context and energy development priorities. For example, developing countries may be driven more by access and equity issues, whereas already developed countries may focus on shifting to renewable energy to meet climate targets.

A new road map will guide major project investments to optimize the outcomes of energy development. Those outcomes include reduced environmental and social risks that could impede energy project finance; energy development mixes that best meet demand and access goals; and climate change mitigation and resilience. To date, upstream planning (done before projects are designed) is rare and does not commonly take into account the spatial footprint of cumulative energy development.[2] Instead, plans are produced at a project level with local considerations about environmental and social impacts as required under environmental impact assessment policies. This is often too late, at too small a scale, and insufficient for addressing the broader environmental risks. When environmental risks are not addressed, the public is likely to rise up in opposition to a project. Growing environmental concerns make it ever more urgent for developers to engage stakeholders early in the process, to sustain public acceptance and to prevent project delays, higher costs, or cancellations (examples of projects stymied by this lack of preplanning include the Inambari Dam in Peru, the Keystone Pipeline in Canada and the United States, and Ontario wind energy projects).[3]

Removing Risk in Energy Financing

Availability of project financing will determine the speed and shape of the energy transition. The financing gap for energy infrastructure is currently estimated at some $28 trillion for 2015 through 2030 (2016 U.S. dollars).[4] This gap can only be closed with access to the capital held by large institutional investors, such as pension, insurance, and sovereign wealth funds. Energy projects are generally considered high risk by investors, and the

main finance challenge is how to derisk them—politically, socially, environmentally—to attract the substantially larger investments necessary to
drive change to a low-carbon future.

Social risks can be considered at the outset of a project by incorporating environmental concerns into landscape-level planning. The typical
approach is instead to site a project and then try to figure out how to mitigate risks.[5] Landscape-level planning helps remove risk from the project
pipeline and promotes pools of bankable projects attractive to affordable
capital. The Organization for Economic Cooperation and Development
has recommended that policy makers drive finance to renewable strategies through advance energy infrastructure planning.[6]

Scenarios for Energy Access

Nearly 1.2 billion people have no access to electricity,[7] and many more need
to consume energy at higher rates in order achieve prosperity.[8] Achieving
full electrification will profoundly impact land use, but this impact can
be reduced if areas that are expanding in power capacity can install new
energy systems in the most effective manner—and, in doing so, leapfrog
ahead of already-developed countries.[9] This can be done by looking at
different energy development scenarios and conducting a comprehensive
risk assessment of the social and environmental impacts associated with
the scenarios (see chapters 4 and 7 for more on scenario planning). When
planning is more predictable and transparent, project developers and communities reduce conflicts, delays, and related costs and accelerate the transition to new energy systems.

Ensuring Development Consistent with Climate
Change Mitigation

Holding the global temperature increase to below 2 degrees Celsius,
as specified by the Paris Agreement, requires severe cuts to emissions
from energy generation. In nearly all cases, this means national energy
portfolios will shift significantly toward renewable sources. Indeed, the
International Energy Agency estimates half the increase in total power
generation between 2015 and 2040 will come from renewables.[10] Other

climate policies, such as a price on carbon or the removal of fossil fuel subsidies, could dramatically change the risk-return profile for renewables investments, strand high-carbon polluting energy infrastructure assets, and spur a more rapid energy transition.[11] A world reliant on renewable energy will give over considerably more land to power generation and transmission in comparison with the amount of land needed by conventional energy sources (see chapters 1, 2, 4, 5, 8, and 9 for more on land-use requirements). These land impacts could be a significant source of greenhouse gas emissions.

The intersection of climate policies, energy development, and emissions due to development-induced land-use change is complex and can only be assessed through landscape planning approaches. Many countries already understand this. The United Nations Food and Agriculture Organization recently reviewed country commitments submitted under the 2015 Paris Climate Change Agreement and found that "land use, land-use change, and forestry" are second only to the energy sector among the most referenced sectors in countries' climate mitigation strategies. Paris commitments also frequently refer to the need for technical support to develop forest inventories and national planning systems, or the need for approaches to reduce dependence on inefficient bioenergy technologies.[12] In addition, implementation of countries' Paris commitments and other climate policies is driving demand for landscape planning to avoid perverse outcomes due to land-use change (e.g., renewable energy development that impacts high-carbon-value areas).

Climate change adaptation policies provide another entry point for integrated planning efforts. Infrastructure projects are themselves highly susceptible to climate change and natural disaster risks[13] that must be factored into planning for sustainability. Landscape-scale planning can help promote project resilience to climate effects, reducing risks for both the project and society.

Pathways to Sustainable Energy Landscapes

There are new approaches to developing sustainable energy, based on next-generation policy and planning approaches that support innovative models for energy development. Below we highlight examples and identify opportunities for improving enabling conditions so that best practice can more rapidly become common practice. Emerging models include:

Solar energy Zones in the United States. As discussed in chapter five, the US Bureau of Land Management adopted a first-of-its-kind Solar Programmatic Environmental Impact Statement in 2012 to accelerate utility-scale solar energy development on public lands by identifying "solar energy zones" while minimizing negative environmental, social, and economic impacts. This landscape plan applies to a six-state region of the southwestern United States and provides an assessment of the likely deployment of solar energy development over the next twenty years and its direct, indirect, and cumulative impacts. To date, nineteen solar energy zones have been identified to support solar energy development, while large areas of the region have also been designated for exclusion from this development because it would not be "the highest and best use of public lands." This approach has encouraged cross-agency collaboration and already reduced the project permitting time by more than half — reducing it to ten months instead of the previous eighteen to twenty-four months.[14]

Sector and Basin Planning for Hydropower in Costa Rica. The Reventazón hydropower project has been developed by the Instituto de Electricidad de Costa Rica, based on a spatially explicit sector plan for hydropower generation and transmission. With participation of the Inter-American Development Bank and the International Finance Corporation in financing the Reventazón project, additional environmental assessments were conducted, including a Cumulative Impact Assessment for the Reventazón basin that identified the most valuable environmental resources and potential mitigation measures.[15] The Reventazón project is now one of the first in Latin America to apply a river offset approach to compensate for project impacts in the basin. The offset is designed to conserve and protect the free-flowing Parismina River and its tributary, the Dos Novillos River, which have natural characteristics similar to the Reventazón.

Landscape-Scale Planning to Guide Sustainable Development in Mongolia. Mongolia's vast landscapes are rich with coal, mineral, and wind energy resources. How this resource development takes place will greatly affect the economy, nomadic livelihoods, and the environment. Recognizing this challenge, the government of Mongolia has supported landscape-scale planning for the entire country to support sustainable development. These plans take into account biological resources, ecosystem services, climate change, and projected development.[16] The plans are helping to guide project siting and mitigation, as well as the establishment of new protected areas that support both nomadic livelihoods and wildlife.

Proactive Planning for Shale Gas in South Africa: The South African government has made high-level commitments to shale gas exploration and, pending successful exploration that yields viable hydrocarbon deposits, the government will likely consider development of those resources at a significant scale. In 2015, the Council for Scientific and Industrial Research partnered with the National Biodiversity Institute and initiated a proactive planning approach to identify development areas while minimizing negative environmental, social, and economic impacts. The planning process traverses Eastern, Northern, and Western Cape provinces and includes twenty-seven local municipalities encompassing 171,811 square kilometers. While this is still a work in progress, it is novel both because of the scale and timing of the assessment. The large scale of the assessment will provide flexibility to assess trade-offs in various alternative development designs and coordinate infrastructure development across multiple developers. By conducting this assessment well in advance of development proposals from industry, regulators can better design concessions that insure environmental and social needs are on equal standing with economic values.

Strengthening the Enabling Conditions for Sustainable Energy Landscapes

Governments, international agencies, financial institutions, industry, and nongovermental organizations all have essential roles to play in strengthening enabling conditions, supporting demonstration models, and institutionalizing new policies and practices for a sustainable energy future. National governments can lead the way by bringing requirements for upstream landscape-scale planning to the energy infrastructure licensing process, and through improvements in land-use and climate policies. They can establish policy, planning, and stakeholder processes that incorporate sustainability incentives and help to de-risk projects for developers and financiers. This requires a robust legal and regulatory framework—with the organizational capacity to back it up—that will support long-term certainty for project finance and environmental and social safeguards.

Multilateral development banks (MDBs) have a tremendous opportunity to drive smarter development by supporting the integration of energy, climate, and biodiversity objectives. MDBs could provide funding and technical support to national governments to conduct landscape-scale, prefeasibility planning for sustainable energy development. This would help address environmental and social risks at the project preparation stage and build stronger pipelines of bankable projects for the MDBs and private financers. Development banks can also continue to lead standard setting for sustainable infrastructure through advancing their own standards, promoting industry standards, building institutional and human resources, and working with governments so that they incorporate sustainability criteria in requests for proposals for energy infrastructure.[17]

Industry and private-sector finance can improve upstream planning for energy infrastructure by promoting strong sustainability standards, providing incentives and resources for landscape-scale planning, and facilitating regional-sector cooperation. Private-sector sustainability standards can provide essential frameworks where governance is weak, especially where they address the full range of sustainability issues[18] and support clear indicators to inform decision making.[19] The private sector can also play an important role by encouraging government-driven efforts to plan in advance for sustainable energy development and by looking for opportunities to facilitate sector cooperation among multiple project proponents for integrated resource corridors and shared infrastructure.[20]

Multilateral institutions such as the International Energy Agency, the International Renewable Energy Agency, and Sustainable Energy for All can support sustainable energy landscapes by building knowledge, capacity, and political support. For example, Sustainable Energy for All's mandate provides an opportunity to promote greater integration of climate, energy access, and sustainability goals into upstream planning. It could convene energy companies, financial institutions, and technical specialists to produce tailored investment plans to address these needs and the upstream planning to achieve them.[21]

Finally, there are multiple ways that nongovernmental organizations and the environmental community can support sustainable energy landscapes. This includes promoting sustainability policies, standards, and practices and supporting landscape-scale planning with technical support to ensure that development avoids impacts to important biodiversity values and ecosystem services. At the same time, the environmental community must acknowledge the goals of sustainable energy landscapes that go beyond conservation, such as meeting energy demand and access goals. Environmental organizations can prevent projects that may have significant environmental impacts and they must help identify go-areas where renewable energy development can occur with less impact.

A sustainable energy future requires that economic, social, and environmental objectives are all addressed in one integrated approach. Society must use the new pathways and solutions emerging for sustainable energy landscapes and intensify efforts to scale and replicate these models of practice. In doing so, more sustainable energy projects will be developed, meeting the broader needs of communities and supporting global sustainability goals.

Notes

1. Gavin Bridge, Stefan Bouzarovski, Michael Bradshaw, Nick Eyre, "Geographies of Energy Transition: Space, Place and the Low-Carbon Economy," *Energy Policy* 53 (2013): 331–40, doi:10.1016/j.enpol.2012.10.066.
2. Anis Radzi and Peter Droege, "Latest Perspectives on Global Renewable Energy Policies," *Current Sustainable/Renewable Energy Reports* 1, no. 3 (2014): 85–93, doi:10.1007/s40518-014-0014-5.

3. John Colton, Kenneth Corscadden, Stewart Fast, Monica Gattinger, Joel Gehman, Martha Hall Findlay, Dylan Morgan, Judith Sayers, Jennifer Winter, and Adonis Yatch-ew, "Energy Projects, Social Licence, Public Acceptance and Regulatory Systems in Canada: A White Paper," *SSRN Journal* (June 2016), doi:10.2139/ssrn.2788022.
4. Aaron Bielenberg, Mike Kerlin, Jeremy Oppenheim, and Melissa Roberts, "Financing Change: How to Mobilize Private-Sector Financing for Sustainable Infrastructure" (McKinsey Center for Business and Environment, January 2016), http://2015.newcli mateeconomy.report/wp-content/uploads/2016/01/Financing_change_How_to _mobilize_private-sector_financing_for_sustainable-_infrastructure.pdf .
5. Helen Moser and Erin Nealer, "Barriers to Bankable Infrastructure" (Center for Strate-gic and International Studies, April 2016), https://csis-prod.s3.amazonaws.com/s3fs -public/publication/160308_Moser_BarriersBankableInfrastructure_Web.pdf.
6. Organisation for Economic Co-operation and Development, "Mapping Channels to Mobilise Institutional Investment in Sustainable Energy," 2015, doi:10.1787/978926 4224582-en.
7. IEA, *World Energy Outlook 2015* (see ch. 1, n. 3).
8. Vaclav Smil, "Global Energy: The Latest Infatuations," *American Scientist* 99, no. 3 (2011): 212–19, doi:10.1511/2011.90.212.
9. Paul Collier and Anthony Venables, "Greening Africa? Technologies, Endowments and the Latecomer Effect," *Energy Economics* 34, supplement 1 (2012): S75–84, doi:10.1016 /j.eneco.2012.08.035.
10. IEA, *World Energy Outlook 2015* (see ch. 1, n. 3).
11. Amar Bhattacharya, Jeremy Oppenheim, and Lord Nicholas Stern, "Driving Sustain-able Development through Better Infrastructure: Key Elements of a Transformation Program," Brookings Global Economy and Development Working Paper no. 91, 2015; N. Stern and P. Calderon, "Better Growth, Better Climate: The New Climate Econo-my Report" (Global Commission on the Economy and Climate, 2014).
12. Food and Agriculture Organization of the United Nations, *The Agriculture Sectors in the Intended Nationally Determined Contributions: Analysis*, June 2016.
13. Gavin Shaw, Jillian Kenny, Arun Kumar, and David Hood, "Sustainable Infrastructure Operations: A Review of Assessment Schemes and Decision Support," in conference proceedings from *25th ARRB Conference,* 2012.
14. Office of NEPA Policy and Compliance, *Final Programmatic Environmental Impact State-ment (PEIS) for Solar Energy Development in Six Southwestern States*, FES 12-24; DOE/ EIS-0403 (Washington, DC: US Bureau of Land Management/US Department of Energy: 2012).
15. Rajesh Singh, H. R. Murty, S. K. Gupta, and A. K. Dikshit, "An Overview of Sustain-ability Assessment Methodologies," *Ecological Indicators* 15, no. 1 (2012): 281–99, doi:10.1016/j.ecolind.2011.01.007.
16. Heiner et al., "Identifying Conservation Priorities" (see ch. 6, n. 8).
17. Bielenberg et al., "Financing Change"; Bhattacharya et al., "Driving Sustainable Development."
18. Bhattacharya et al., "Driving Sustainable Development"; Joerg Hartmann, David Har-rison, Jeff Opperman, and Roger Gill, Jonathan Higgins, Amy Newsock, Randy Curtis, Denny Grossman, Paulo Petry, Rosario Gomez, Juliana Delgado, Tomas Walsch-burger, Jose Yunis, Juan Carlos Gonzalez, Ana Cristina Barros, and Nelida Barajas,

"The New Frontier of Hydropower Sustainability: Planning at the System Scale" (The Nature Conservancy, 2013); Heiner et al., "Identifying Conservation Priorities"; Cesar Poveda and Michael Lipsett, "A Review of Sustainability Assessment and Sustainability/Environmental Rating Systems and Credit Weighting Tools," *Journal of Sustainable Development* 4, no. 6 (2011), doi:10.5539/jsd.v4n6p36; Jianguo Wu and Tong Wu, "Sustainability Indicators and Indices: An Overview. In *Handbook of Sustainable Management*, edited by Christian Madu and Chu-Hua Kuei (London: Imperial College Press, 2012); International Finance Corporation, *IFC Performance Standards on Environmental and Society Sustainability*, 2012, http://www.ifc.org/wps/wcm/connect/c8f524004a73 daeca09afdf998895a12/IFC_Performance_Standards.pdf?MOD=AJPERES.

19. Spiro Pollalis, Andreas Georgoulias, Stephen Ramos, and Daniel Schodek, eds., *Infrastructure Sustainability and Design* (New York: Routledge, 2012); Georgoulias Andreas, Jill Allen, Libby Farley, Jon Kao, and Irina Mladenova, "Towards the Development of a Rating System for Sustainable Infrastructure: A Checklist or a Decision-Making Tool?" from *Proceedings of the Water Environment Federation, Cities of the Future/Urban River Restoration* 13 (2010): 379–91, doi:10.2175/193864710798284643.

20. Adam Smith International, "Integrated Resource Corridors Initiative: Scoping and Business Plan," 2015.

21. Carlos Pascual and Jason Bordoff, "A Global Low-Carbon Challenge," *Democracy: A Journal of Ideas* (Winter 2016), no. 39, http://democracyjournal.org/magazine/39 /a-global-low -carbon-challenge/.

CHAPTER 11

The Last Word

Joseph M. Kiesecker and David E. Naugle

Wrapping up, we reflect on our three primary motivations for writing this book:

1. Heighten as a global issue our expanding energy footprint and its potential future impacts on people and nature.
2. Provide solutions to the biodiversity crisis given trade-offs that accompany a rapid shift to renewable energy to thwart climate change.
3. Introduce tools in hand that, if applied broadly, can put us on a path to sustainability.

There are six things we can do today to better balance global development and conservation:

• Increase Urgency

Planning for our own sustainability in a piecemeal fashion lacks the urgency necessary to solve this issue. We hope this book helps society think about energy as a single unifying issue that must be wholly addressed. Despite our daily reliance, people by their very nature are creatures of crisis, not worrying about where food and energy comes from until they become scarce. Climate aside, the relatively small footprint of conventional energy sources in part explain the lack of urgency—but this is changing now as we transition to more renewable energy sources and potential conflicts over land use could become more likely.

• **Accept Trade-offs from Renewables**

The human footprint required to generate energy from renewables exceeds that of traditional carbon sources. Now that we have acknowledged this, we can address it. Likewise, we hope this book enables people to shed their denial or guilt about elevated impacts resulting from society's shift to renewable energy. Unless this happens, we will never reduce the far greater impacts of climate change.

• **Reduce Time Lags**

Society always seems slow to change, but then change seems to happen in an instant—and we hope our shift to renewables to soften climate impacts is no exception. But we have to act quickly as the seeds of the next energy century—power plants, refineries, and other energy infrastructure—are being sown as you read, and their long life spans influence society's ability to change. It's also clear that the world has a huge investment in the current energy infrastructure and it is simply not going to write it off. But that doesn't mean we have to wait to implement the renewable energy solutions we have now. Every step in the wrong direction makes it less likely we can alter the trajectory of a rapidly changing climate.

• **Facilitate Master Planning**

A primary and noticeable outcome of change is when best practices become common practices. But we are still perplexed that the best practices we have put forward in part II are still not more widely embraced. Imagine you want to build your new dream home—you have thought out every detail, from the modern kitchen to built-in library bookshelves. Your contractor initiates work without a floor plan, instead deciding to figure it out as they go. No homeowner in their right mind would allow this, but this is exactly how we are developing natural landscapes all over the world—with no master plan or vision to guide us. Landscapes need a vision for the same reason that builders need a blueprint—to do it right. Changing course before work begins is cheap and fast while trying to fix mistakes afterward is expensive, slow, and usually impossible to achieve.

- **Policy and Enabling Conditions**

Achieving sustainability hinges in part on our ability to break down long-held regulatory and financial barriers that enable master planning. Society needs more examples of pathways that help facilitate this transition. This should include a library of practices and policies that countries can adapt or insert into national level policy. The environmental community should help simplify this for countries willing to make necessary change.

- **Providing Next-Generation Proof of Concept**

We encourage more case studies showing how multiobjective planning across sectors can provide social and economic benefits to developing countries rich in natural resources. Developed countries have destroyed much biodiversity and are largely responsible for climate change. Asking developing countries to forgo quality of life to abate environmental crises smacks of ecoimperialism and is unjust unless we plan to compensate their decision to forgo development. Conservation will remain a luxury until we achieve development in ways that allow developing countries to improve their well-being while providing for nature.

About the Editors

Joseph M. Kiesecker is Lead Scientist for The Nature Conservancy, the world's largest environmental organization. He is a recognized scientific leader in biodiversity conservation and in developing guidance for application of the mitigation hierarchy. He has also been at the forefront of moving development and mitigation planning to a landscape scale. He pioneered the Conservancy's Development by Design approach, an initiative that seeks to improve development planning through the incorporation of predictive modeling to provide solutions that benefit conservation goals and development objectives and directs its implementation in over a dozen countries around the world. To date, the program has helped shape changes to national level environmental licensing policy in the United States, Mongolia, and Colombia; helped to guide the implementation of approximately $1 billion in offset and other conservation funding in the United States; and helped guide the establishment of approximately 150,000 square kilometers of new protected areas in Mongolia.

His past work has focused primarily on the conservation and ecology of freshwater systems. In particular, he has been interested in the global amphibian decline phenomenon. This line of research has involved investigating how perturbations resulting from climate change and land use changes can stress organisms, making them more susceptible to disease. Joe has published over 130 articles, on topics ranging from climate change to the effectiveness of conservation strategies; examples of his work have been published in *Nature, Science, Proceedings of the National Academy of Sciences, Conservation Biology, Ecology,* and *American Scientist.* Prior to joining The Nature Conservancy, Joe was a professor at Penn State University and has also held faculty appointments at Yale University and the University of Wyoming.

David E. Naugle is a professor in the Wildlife Biology program at the University of Montana, Missoula. His applied science emphasizes biological planning and outcome-based evaluations in landscape conservation. Seventeen years spent investigating energy and wildlife issues culminated in his 2011 book with Island Press, *Energy Development and Wildlife Conservation in Western North America*. Dave has published over 100 articles, ranging widely from disease to conservation policy; findings of his work have appeared in *Science, BioScience, Conservation Biology, Ecology Letters*, and *PLosOne*.

In his former life, Dave mostly fought industry, but he now seeks innovative solutions to balance energy and biodiversity futures—the focus of this new book. Since 2010, Dave also has served as US Department of Agriculture's Science Adviser to the Natural Resources Conservation Service–led Sage Grouse Initiative, where he helps guide a $0.5 billion Farm Bill investment to achieve wildlife conservation through sustainable ranching. The Sage Grouse Initiative has emerged as a primary catalyst for voluntary and incentive-based conservation on private lands that checkerboard the western United States. With more than 1,450 participating ranches in eleven western states, the Initiative and its partners have conserved more than five million acres through highly targeted investments. In 2016, Secretary of Agriculture Tom Vilsack selected Dave and his colleagues to receive USDA's Abraham Lincoln Award for innovations in conservation.

Contributors

Amal-Lee Amin is Chief of the Climate Change division at the Inter-American Development Bank, where she manages an extensive portfolio of climate change adaptation, mitigation, and knowledge activities for all of Latin America and the Caribbean. Prior joining the Bank, Amal-Lee was Associate Director at E3G, leading a program on international climate finance. Her in-country work focused on design and implementation of financing strategies for low-carbon and resilient development in Latin America, Africa, and Asia. From 2001 to 2011, Amal-Lee worked for the United Kingdom and European Union governments, developing their policies for climate and sustainable energy. Achievements include the design of a new Green Investment Bank for the United Kingdom, developing strategy for negotiations under the United Nations Framework on Climate Change, the United Nations General Assembly, and the 2002 World Summit on Sustainable Development ("Rio+10"). Amal-Lee's PhD research, which focused on policies and institutions for increasing investment in renewable energy in developing countries, has been central to her fifteen-year professional career.

Sharon Baruch-Mordo is a spatial scientist on the Global Lands Science team at The Nature Conservancy. As a quantitative ecologist, she enjoys applying quantitative methods in ecology to conservation issues on local and global scales. Her research interests are varied and include spatial ecology, conservation biology, and energy development. Her past research focused on greater sage-grouse conservation as part of ongoing collaboration between The Nature Conservancy and the Sage Grouse Initiative. Her work on sage-grouse sought to develop range-wide models to better understand effects of conservation threats on sage-grouse demographics in order to predict change under differing landscape and management scenarios. Before going grouse, Sharon's research focused on urban black bear ecology and human-bear interactions and conflicts. She completed her PhD degree in Ecology (2012) and an MS degree in Wildlife Biology (2007) at the Department of Fish, Wildlife, and Conservation Biology at Colorado State University.

Leandro Baumgarten is an ecologist with a PhD degree in ecology from the University of Campinas in Brazil. His experience includes landscape ecology and systematic conservation planning. He started his career in a conservation project on one of the most endangered birds of prey in Brazil, the crowned eagle. He has worked with nongovernmental organizations and environmental state agencies in Brazil, coordinating the revision of the priority areas for conservation of the Atlantic Forest for the Brazil Ministry of the Environment before joining The Nature Conservancy in 2007, first as the Science Coordinator of the Central Savannas and now as the Science Manager for the Brazil Conservation Program. Currently, he also coordinates the Transparency Strategy within the Collaboration for Forests and Agriculture (the Gordon and Betty Moore Foundation, The Nature Conservancy, the World Wildlife Fund, and the National Wildlife Federation). This initiative has the bold goal of removing deforestation from the supply chains of beef and soy in the Amazon, Cerrado, and Chaco.

D. Richard Cameron is an Associate Director of Science for the California chapter of The Nature Conservancy, managing the Land Program science team. His work at the Conservancy is focused on creating the tools and evidence to integrate conservation into land-use and climate policies. Across a variety of landscapes and themes, his research assesses the potential alignment between conservation objectives and other societal goals, such as alternative energy development, transportation infrastructure, food production, and climate stabilization. Previously, he worked for GreenInfo Network, where he specialized in helping organizations and public agencies design and communicate strategic priorities. His academic background is in geography, with a BA degree from Middlebury College and an MA degree from the University of Colorado.

Juan José Cárdenas is an oceanographer (Univ. Bretagne Occidentale, France) with an MS degree in fisheries and agriculture (Univ, Bretagne Occidentale—Pièrre et Marie Curie Paris VI, France). He is the Environmental and Fisheries Project Coordinator of the Fundación para la Pesca Sostenible y Responsable de Túnidos (FUNDATÚN), Caracas, Venezuela, and was formerly the Fisheries and Infrastructure Marine Conservation Coordinator for the Northern Tropical Andes and Southern Central America for The Nature Conservancy. His areas of expertise include fisheries acoustics, design and implementation of management systems focused on planning and execution of environmental conservation guidelines, and management of technical and educational research centers. He has participated in numerous environmental impact assessments, biological baseline studies, and he is author, editor, and/or coauthor of more than twenty-five technical papers.

Laura Crane is the Deputy Director of the Land Program for the California chapter of The Nature Conservancy. She oversees a body of work that is focused on achieving conservation outcomes by influencing land use and energy planning, permitting, and policies. In addition, she oversees the Conservancy's program to improve the ecological resilience of California's Islands. Prior joining the Conservancy in 2003, Laura was the Director of the California Renewable Energy Initiative. In this role, she provided scientific analyses and collaborated with government, industry, and other nongovernmental organizations to develop solutions for preserving and protecting unique biodiversity and ecosystem function while also facilitating low-impact renewable energy development. She also spent nine years as an environmental and resource management consultant, developing soil, river, and groundwater restoration plans for contaminated Superfund sites. She holds a BA degree in environmental studies from the University of California, Santa Barbara.

Kevin Doherty is a spatial ecologist for the U.S. Fish and Wildlife Service. In this position, he is responsible for creation and integration of applied conservation research that addresses conservation issues for migratory birds and other priority species. Doherty has published forty scientific peer-reviewed papers and book chapters on topics including landscape ecology, sage-grouse and sagebrush ecology, spatial ecology of water-fowl, movement ecology, as well as conservation planning and mitigation policy in relation to energy development. His expertise in geographic information system—based habitat modeling and landscape ecology has influenced conservation policy and land management decisions by providing the scientific basis on which multistakeholder groups have developed plans to balance conservation with other land uses.

Jeffrey S. Evans is a landscape ecologist with The Nature Conservancy and an Affiliate Assistant Professor with University of Wyoming, Department of Zoology and Physiology. His focus is on the development of statistical tools for evaluation of species status, biodiversity, habitat quality, and potential effects of anthropogenic development. Working extensively on the Conservancy's Development by Design strategy, he leverages disciplines such as ecology, spatial statistics, genetics, and remote sensing to develop solutions that benefit conservation outcomes. Jeffrey has provided statistical technical consultation to several US federal agencies and, to date, has published more than eighty peer-reviewed journal articles. Prior to joining The Nature Conservancy, Jeffrey worked for the USDA Forest Service in policy, management, and research roles.

Joe Fargione is Science Director for The Nature Conservancy's North America Region. Joe's research seeks ways to balance human energy and food demands with environmental conservation. Solutions include appropriate siting of new energy development and new sources for conservation funding, including natural infrastructure, compensatory mitigation payments, carbon offsets, and creating markets that value nature's benefits. Prior to joining the Conservancy, Joe received a bachelor's degree from Hampshire College and a PhD degree in ecology from the University of Minnesota and held faculty positions at the University of New Mexico and Purdue University. Joe's dozens of scientific publications have been cited thousands of times and have generated national media coverage, including by *NBC Nightly News*, *Time Magazine*, the *New York Times*, and the *Onion*, among others. Joe is a native of Minneapolis, Minnesota, where he resides with his wife and children.

Juan Carlos González is an economist with a master's degree in environmental planning. From 2004 until 2016 he was the Coordinator of Infrastructure and Biodiversity in Northern Andes and South Central America at The Nature Conservancy. During that time he participated in environmental planning projects with Petróleos de Venezuela, S.A. in Venezuela and the development of a freshwater portfolio for the Orinoco basin in Colombia with the Agencia Nacional de Hidrocarburos and the Ministry of Environment. In the Peruvian Amazon, he participated in an environmental planning process for new roads and transmission line projects. In Ecuador, he was part of an expert group that developed a portfolio for conservation in the northern Amazonia. He was one of the authors of the Oil and Gas Offset Guidelines (Global Reporting Initiative), resulting in a change of the national offset regulations in Colombia. At present he is an independent consultant, and as such has participated in the new Environmental Plan for Quito. He has also been an adjunct professor for the Green Infrastructure Program at the Universidad Javeriana in Colombia and in the Universidad Católica de Ecuador.

Peter L. Hawthorne is a Senior Scientist with the Natural Capital Project, a partnership between Stanford University, the University of Minnesota, The Nature Conservancy, and the World Wildlife Fund. Peter's work focuses on resolving the challenge of meeting multiple objectives in ecosystem services planning. To tackle this problem, he develops methods and tools that help visualize, prioritize, and optimize across multiple services and planning targets, with the goal to make this information accessible and interactive for use in decision making. He enjoys that this work continually draws on his background in mathematics, ecology, economics, and even philosophy. In addition to his PhD degree in ecology, evolution and behavior from the University of Minnesota, Peter holds an AB degree in mathematics from Harvard University.

Mark Hebblewhite is a professor of ungulate ecology in the Wildlife Biology Program at the University of Montana in Missoula. He has conducted research on large carnivores and their ungulate prey since 1994 across Canada, Europe, and Asia. Mark's research uses strong empirical approaches to the study of human impacts on wildlife to develop effective conservation strategies for large mammals and the ecosystems in which they live. Mark uses large ungulates and their predators as good entry points to understanding ecosystems because of their important roles and their conservation relevance. Mark served on the Federal Environment Canada Critical Habitat Science Team from 2007 to 2011, a team that identified critical habitat for boreal woodland caribou over 1,000,000 square kilometers of Canada.

Christina M. Kennedy is a Senior Scientist for the Global Lands Program at The Nature Conservancy. In this role, she leads science initiatives to help reconcile human land use with nature conservation. She works with public agencies, corporations, civil society groups, and research institutions to integrate the best science and tools on conservation planning, landscape ecology, agroecology, and impact mitigation to promote sustainable landscape design. Christina's research examines the effects of land use, landscape pattern, and habitat fragmentation on species, ecosystems, and the services they provide. Her projects span local to global scales and integrate field studies, landscape modeling, and data synthesis with the aim of improving land use practice and policy. Christina's training is in conservation biology and landscape ecology, with a PhD degree from the University of Maryland, a master's degree from Duke University, and a BS degree from Cornell University.

Eduardo Klein is a marine biologist (Univ. Simón Bolívar, Venezuela) with a master's degree in marine resources management (Univ. Quebec, Canada). He is an Associate Professor in the Environmental Studies Department and a researcher at the Institute of Marine Science and Technology and the Center for Marine Biodiversity at the Universidad Simón Bolívar in Caracas, Venezuela. His areas of expertise include the ecology of coastal marine fishes, the use of remote sensing tools for analysis of coastal and oceanographic processes, and spatial modeling for conservation of marine biodiversity. He is member of the Ocean Biogeographic Information System steering group and is a resource person for the Marine Program of the Convention for Biological Diversity. He has participated in more than thirty environmental impact assessments, biological base-line studies, and risk analyses for the oil and gas industry.

Gert Jan Kramer is Professor of Sustainable Energy Supply Systems at the Copernicus Institute of Sustainable Development of Utrecht University, the Netherlands. His research interests concern both the technological and the societal aspects of the energy transition. His research activities include technology assessment, energy scenarios, and actor-based modeling of the energy system. Prior to taking up his current position, he worked for twenty-seven years at Shell, mostly in alternative energy research. The last six years of his tenure at Shell he led a small think tank on energy futures, working closely with Shell's scenario team. He is the editor of a recent collection of essays on the future of energy in society, *The Colours of Energy*, available at www.shell.com/colours.

Linda Krueger is Senior Policy Advisor at The Nature Conservancy, where she has been responsible for integrating mitigation methods into national regulatory policies and for advancing mitigation principles among global financial institutions and nongovermental organizations. Before joining the Conservancy, she was vice president at the Wildlife Conservation Society, where she directed government and multilateral agency relations through offices in New York, Washington, DC, and Brussels. As an Adjunct Professor at Columbia University's School of International and Public Affairs, she taught a graduate course on global environmental politics from 2008 to 2012. Previously she served as foreign policy legislative staff in both the United States Senate and the House of Representatives and as a consultant to the North Atlantic Treaty Organization Scientific and Environmental Affairs Division. She holds both undergraduate and graduate degrees from Stanford University.

Roger Martínez is an Urban Planner (Univ. Simón Bolívar, Venezuela) with a master's degree in systems engineering (Univ. Simón Bolívar, Venezuela), and a PhD degree in architecture (Univ. Central de Venezuela). He is a Titular Professor of the Urban Planning Department and a research collaborator at the Institute of Marine Science and Technology of the Universidad Simón Bolívar in Caracas, Venezuela. His field of work is in the areas of social sciences, environmental and urban planning, and sanitary infrastructure planning. He is a member of the Environmental Committee of the Venezuelan Academy of Physics, Mathematics, and Natural Sciences. His expertise as a consultant in social and environmental matters includes environmental impact assessments and baseline studies for the oil and gas industry, as well as water and sanitation studies and urban planning.

Bruce McKenney is Director for Development by Design at The Nature Conservancy, where he leads a global team to advance solutions for conservation and responsible energy, mining, and infrastructure development. He has spent more than twenty-five years working at the intersection of development and environmental challenges. Prior to joining the Conservancy, Bruce directed the Natural Resources and Environment program at the Cambodia Development Resource Institute. Over his career, he has conducted projects and studies for the World Bank, the US Fish and Wildlife Service, the World Wildlife Fund, and the World Commission on Dams. Bruce has served on the World Economic Forum Global Agenda Council and has published more than a dozen journal articles and book chapters. He holds a master's degree in public policy from Harvard University, an honors degree in African studies from the University of Cape Town, and a BA degree with honors in political science from Brown University.

Daniela A. Miteva is an environmental economist working primarily on conservation and sustainability issues in developing countries. Combining a microeconomic framework with theory and tools from ecology and biogeography, her research focuses on understanding the drivers of landscape change, quantifying the impacts on ecosystems and human welfare, and evaluating policies like protected areas and Forest Sustainability Council certification. Daniela is currently an Assistant Professor of sustainable development and economy at Ohio State University. She holds a PhD degree from Duke University and a dual BA degree in biology and economics from Bryn Mawr College.

James Oakleaf is part of The Nature Conservancy's Global Lands Science Team in Fort Collins, Colorado. Within this team, Jim leads or assists with applying visualization techniques, using open-source geospatial technologies and data, developing web mapping and desktop applications, integrating geographic information systems with environmental modeling, and global/regional conservation planning. His work recently has centered on creating tools in both Brazil and Mongolia to support country-wide environmental regulations. He also is currently coleading efforts by the Nature Conservancy to globally assess potential development threats and future land conversion in order to aid the Conservancy in prioritizing future conservation efforts across our planet. Prior to joining the Conservancy, Jim worked at the University of Wyoming teaching geospatial technologies and applying these technologies to natural resources issues in Wyoming and across the Rocky Mountain region.

Jeff J. Opperman is the Lead Scientist for The Nature Conservancy's Great Rivers Program. Jeff has been working to protect rivers and floodplains for more than fifteen years, collaborating with teams in the United States, Asia, Africa, and Latin America on river conservation projects and research. His expertise includes floodplain ecology, environmental flows, and the integration of river conservation with sustainable hydropower. Jeff earned his BS degree in biology from Duke University and a PhD degree in ecosystem science from the University of California, Berkeley, and then studied floodplain ecology during a postdoctoral fellowship at the University of California, Davis. His scientific and policy research has been published in journals such as *Science*, *BioScience*, and *Ecological Applications*. Jeff strives to communicate the challenges and opportunities of protecting freshwater through op-eds, articles, and blog posts in such places as the *New York Times*, *Outside*, and the *Guardian*.

Juan Papadakis is a biologist (Univ. Simón Bolívar, Venezuela), with thirteen years of experience creating, developing, and maintaining geographic information systems for data storage and digital mapping. He has participated in projects for the selection of priority conservation areas, sustainable use of natural resources, environmental impact assessments, and biological baseline studies in the Southern Caribbean and the Orinoco Basin in Venezuela.

Sophie S. Parker is a Senior Scientist in the Los Angeles office of The Nature Conservancy. She has provided scientific leadership to Conservancy projects in California since 2008 and has focused on protecting ecologically important lands and waters in the southern part of the state. Some of Parker's current endeavors include serving as the Science Lead for the California Renewable Energy Program, the restoration of the Santa Clara River in Ventura County, and the Greater Los Angeles Urban Conservation Program. Parker received her PhD degree from the Department of Ecology, Evolution, and Marine Biology at the University of California, Santa Barbara, in 2006. While in Santa Barbara, she focused her research belowground, examining how soil nitrogen and symbiotic fungi prevent the reestablishment of native bunchgrasses in previously invaded California grasslands. One of Parker's long-term goals is to better integrate the fields of soil science and ecosystem ecology into conservation practice.

John M. Randall is a Lead Scientist for The Nature Conservancy's California chapter. He leads a team of four other scientists with a focus on biodiversity conservation in protected area networks, including the linkages between core conservation lands and waters. John led the team that conducted the Conservancy's 2010 Mojave Desert Ecoregional Assessment. His current work includes research on impacts of climate change and conservation management on the California Channel Islands, connectivity and enhancing climate adaptation of conservation reserve networks across mainland Southern California, and urban biodiversity conservation across Greater Los Angeles. John joined The Nature Conservancy in 1991 as an invasive species specialist, became the organization's global invasive species program director in 2004, and moved to the Conservancy's California chapter in San Diego in May 2009.

Kei Sochi is a spatial ecologist with the Global Lands Science team at The Nature Conservancy and has worked on projects examining the impacts of development in Australia, Indonesia, Brazil, and the United

States. Before joining the Global Lands team, Sochi worked for the Conservancy's Nevada and Colorado field offices on regional, statewide, and local conservation planning and analysis efforts. Kei received her AB degree from Princeton University and an MA degree from the University of Pennsylvania.

Heather Tallis is the Chief Scientist for Strategy Innovation for The Nature Conservancy. Throughout her career, she has been a recognized scientific leader in ecosystem services and in developing guidance for the mitigation hierarchy, among other practical applications of ecosystem service science. She joined the Conservancy in 2013 as the organization's first female lead scientist. She previously was Lead Scientist for the Natural Capital Project, a partnership between Stanford University, the University of Minnesota, The Nature Conservancy, and the World Wildlife Fund, where she led the development of InVest, a software application that reveals the ecosystem service costs and benefits of land- and water-use decisions. Tallis also supported development of OPAL, a tool for assessing biodiversity and ecosystem services in development and mitigation planning. She is coauthor of the National Ecosystem Services Partnership's Best Practices for Integrating Ecosystem Services into Federal Decision Making. Tallis is a leading innovator in bringing human well-being considerations into conservation and has developed many research programs with The Nature Conservancy that help understand and manage key connections between humans and nature.

Elizabeth M. Uhlhorn serves as Dow's Sustainability Program Manager for Ecosystem Services. In this role, she is serving as project manager for implementation of Dow's new 2025 Valuing Nature Goal, which aims to generate $1 billion in business value through projects that are good for the environment. Uhlhorn also manages Dow's $10 million collaboration with The Nature Conservancy. Prior to joining the sustainability team, she held several finance positions throughout Dow. Before joining Dow, Uhlhorn worked at the World Wildlife Fund. She has also served in the US Peace Corps in Cameroon, where she trained farmers in natural fertilization techniques and developed small agricultural businesses. Uhlhorn holds dual MBA and MS degrees through the Erb Institute for Global Sustainable Enterprise at the University of Michigan. She also received a BA degree in environmental science and political science from the University of St. Thomas, in St Paul, Minnesota.

Graham Watkins is a Principal Environmental Specialist in the Climate Change and Sustainability Division of the Inter-American Development Bank. For the past six years, he has been a member of the environmental and social safeguards management team for the same bank. He has over twenty years of experience in stakeholder involvement, field research, and policy development, including periods as the Executive Director of the Charles Darwin Foundation in Galápagos and the Director General of the Iwokrama International Centre for Rain Forest Conservation and Development in Guyana. He has worked extensively in biodiversity research, natural resources management, environmental and social assessment, and sustainable infrastructure and across multiple stakeholder groups in the collaborative development of plans for protected areas and institutional development. Graham has a PhD degree in ecology and evolution from the University of Pennsylvania in Philadelphia and a BA degree in zoology from the University of Oxford.

Acknowledgments

First and foremost we thank our chapter authors, who gave freely of their time and expertise. It was a pleasure to work with such a talented group of scientists and conservation practitioners. Their passion for both the topic and this project made our job fun.

Erin Johnson and the team at Island Press were amazing all the way— from initial conversations through the push to final print. Leana Schelvan's stylistic edits made us better storytellers; and Emily Harrington's graphics made the pages come alive! For your teamwork and expertise we are grateful.

Our former academic lives were spent documenting impacts from development—this book is a culmination in our collective desire to provide innovative solutions to balance energy development and conservation. This shift was guided by our many colleagues working on the front lines of conservation at The Nature Conservancy, the US Bureau of Land Management, and the US Department of Agriculture's Natural Resources Conservation Service. We are most grateful to Jeff Evans, Kei Sochi, Mike Heiner, Jim Oakleaf, Christina Kennedy, Sharon Baruch-Mordo, Holly Copeland, Amy Pocewicz, Andrea Erickson, Bruce McKenney, Tomas Walschburger, Aurelio Ramos, Joe Fargione, Rob McDonald, Davaa Galbadrakh, Oidov Enkhtuya, Roy Vagelos, Nels Johnson, Judy Dunscomb, Thomas Minney, and James Fitzsimons at The Nature Conservancy; colleagues on the Conservancy's Global Lands and Development by Design teams always challenged us to do good science and to make the resulting products relevant to managers and decision makers. Special thanks to Peter Kareiva and Justin Adams for giving Joe the time and support to accomplish this project.

Much of Joe's early thinking was influenced by colleagues working in The Nature Conservancy's regional and country programs around the world who through their understanding of the local contexts helped shape his understanding of energy sprawl as a global issue and how best to address it. Of note are the Conservancy's Wyoming and Colorado chapters, who were there from the start, and early adopters like The Nature Conservancy in Australia, Colombia, Indonesia, Mongolia, and Zambia; they really helped him understand how to adapt his thinking to international contexts and to feel the urgency for solutions in developing countries. He is also particularly grateful to government partners in these countries, in particular the Colombian Ministry of Environment and Sustainable Development and the Mongolian Ministry of Environment and Green Development.

Much of Dave's early thinking was shaped by Bureau of Land Management colleagues engaged in energy development on public lands. Together, folks like Dale Tribby, John Carlson, and Tom Rinkes introduced Dave to oil and gas in the North American West and all the policy decisions that accompany it. More recently, Tim Griffiths, Jeremy Maestas, and Thad Heater, all part of the Sage Grouse Initiative led by the Natural Resources Conservation Service, opened his eyes to wildlife conservation through sustainable ranching in the sagebrush steppe, a success story highlighted in chapter 3. For flexibility to work on this book Dave thanks his University of Montana administrators, including now-retired Provost Perry Brown, previous Director of the Wildlife Biology Program Daniel Pletscher, and current supervisor Chad Bishop.

We also thank the numerous organizations that have funded our work in this field. Of note are the Robertson Foundation, the 3M Foundation, and the Anne Ray Charitable Trust; the Mongolian and Colombian governments, the US Bureau of Land Management, the US Department of the Interior, the US Fish and Wildlife Service, and the Natural Resources Conservation Service. We also appreciate the wicked problems our corporate partners asked us to help solve. These situations forced us to consider options to better balance both conservation goals and development objectives; notable are engagements with British Petroleum, Shell, Rio Tinto, Barrick Gold, QEP Resources, and Dow.

The urgency portrayed throughout this work comes in large part from personally experiencing the forthcoming energy transformation—seeing firsthand the people and places at risk makes it all too real. But those experiences come with sacrifices—namely time away from our families. So we are thankful to them for their support, understanding, and encouragement. Special thanks to Cheri, Jackson, and Griffin Kiesecker, who are always willing to be the Clark to Joe's Lewis; and to Corey, Madelyn, and Hunter Naugle for gallivanting far and wide.

Joseph M. Kiesecker, Fort Collins, Colorado, January 2017
David E. Naugle, Missoula, Montana, January 2017

Index

Page references followed by an "f" indicate figures/photos/illustrations; Page references followed by a "t" indicate tables.

About Island Press

Since 1984, the nonprofit organization Island Press has been stimulating, shaping, and communicating ideas that are essential for solving environmental problems worldwide. With more than 1,000 titles in print and some 30 new releases each year, we are the nation's leading publisher on environmental issues. We identify innovative thinkers and emerging trends in the environmental field. We work with world-renowned experts and authors to develop cross-disciplinary solutions to environmental challenges.

Island Press designs and executes educational campaigns in conjunction with our authors to communicate their critical messages in print, in person, and online using the latest technologies, innovative programs, and the media. Our goal is to reach targeted audiences—scientists, policymakers, environmental advocates, urban planners, the media, and concerned citizens—with information that can be used to create the framework for long-term ecological health and human well-being.

Island Press gratefully acknowledges major support of our work by The Agua Fund, The Andrew W. Mellon Foundation, The Bobolink Foundation, The Curtis and Edith Munson Foundation, Forrest C. and Frances H. Lattner Foundation, The JPB Foundation, The Kresge Foundation, The Oram Foundation, Inc., The Overbrook Foundation, The S.D. Bechtel, Jr. Foundation, The Summit Charitable Foundation, Inc., and many other generous supporters.

The opinions expressed in this book are those of the author(s) and do not necessarily reflect the views of our supporters.